G. Seminara P. Blondeaux (Eds.)

River, Coastal and Estuarine Morphodynamics

Springer

Berlin
Heidelberg
New York
Barcelona
Hong Kong
London
Milan
Paris
Singapore
Tokyo

G. Seminara P. Blondeaux (Eds.)

River, Coastal and Estuarine Morphodynamics

With 111 Figures and 2 Tables

Springer

PROFESSOR GIOVANNI SEMINARA
University of Genova
Department of Environmental Engineering
Via Montallegro 1
16145 Genova
Italy

PROFESSOR PAOLO BLONDEAUX
University of Genova
Department of Environmental Engineering
Via Montallegro 1
16145 Genova
Italy

ISBN 3-540-41839-3 0 Springer-Verlag Berlin Heidelberg New York

Library of Congress Cataloging –in-Publication Data
River, coastal and estuarine morphodynamcs / Giovanni Seminara; Paolo Blondeaux. p.cm. "This book collects review papers on most of the topics covererd by the invited lecturers at the IAHR Symposium on River, Coastal, and Estuarine Morphodynamics, held in Genova (Italy) on September 6th-10th, 1999"—Pref.
Includes bibliographical references.
ISBN 3-540-41839-3

Springer-Verlag Berlin Heidelberg New York
a member of BertelsmannSpringer Science+Business Media GmbH

http:// www.springer.de

Printed in Germany

Cover design: Erich Kirchner, Heidelberg
Typesetting: camera-ready by the editors
Printed on acid-free paper SPIN 10831330 32/3130/as 5 4 3 2 1 0

Preface

This book collects review papers on most of the topics covered by the invited lecturers at the IAHR Symposium on River, Coastal and Estuarine Morphodynamics, held in Genova (Italy) on September 6th-10th, 1999. The above meeting was conceived and organized with the aim to emphasize the unity of the subject of morphodynamics and the role played by cross-fertilization in its recent development.

The publication of this book has been possible thanks to the efforts of several scientists who have accepted to act as referees and helped us reviewing the submitted papers. Let us thank Aronne Armanini, Rolf Deigaard, Gianpaolo Di Silvio, Rob Ettema, David Furbish, Marcelo Garcia, Yarko Niño, Joshua Roering, Yasuharu Shimizu, John Sleath, Marcel Stive, Ed Thornton, Marco Tubino, Huib De Vriend and Guido Zolezzi.

Grateful thanks are also due to Tina d'Agostino who has done a considerable amount of editorial work and to Maurizio Brocchini who has managed the whole process.

Along with the Proceedings of the IAHR Symposium, the present volume represents one of the outcomes of the activities of the Italian research group 'Morfodinamica Fluviale e Costiera', coordinated by G. Seminara and cofunded by the Italian Ministry of Scientific Research and by various Italian Universities.

Genova,
February 2001

Giovanni Seminara
Paolo Blondeaux

Contents

List of Contributors

Paolo Blondeaux, Dipartimento di Ingegneria Ambientale, Università di Genova, Via Montallegro 1, 16145 Genova, ITALY

Rolf Deigaard, Department of Hydrodynamics and Water Resources (ISVA), Building 115, Technical University of Denmark, DK-2800 Lyngby, DENMARK

Huib De Vriend, Faculty of Civil Engineering & Geosciences, Delft University of Technology, POB 5048, 2600 GA, Delft, NL

Juan J. Fedele, V.T. Chow Hydrosystems Laboratory Department of Civil and Environmental Engineering, University of Illinois at Urbana-Champaign, USA

Jørgen Fredsøe, Department of Hydrodynamics and Water Resources (ISVA), Building 115, Technical University of Denmark, DK-2800 Lyngby, DENMARK

Marcelo H. García, V.T. Chow Hydrosystems Laboratory Department of Civil and Environmental Engineering, University of Illinois at Urbana-Champaign, USA

Rob Holman, Oregon State University, Corvallis, Oregon, 97331-550, USA

Jasmin Imran, Department of Civil and Environmental Engineering University of South Carolina, USA

Stefano Lanzoni, Dipartimento di Ingegneria Idraulica, Marittima e Geotecnica, Università di Padova, via Loredan 20, 35131 Padova, ITALY

Marco Marani, Dipartimento di Ingegneria Idraulica, Marittima e Geotecnica, Università di Padova, via Loredan 20, 35131 Padova, ITALY

Stephen R. McLean, Department of Ocean Engineering, UCSB, Santa Barbara, California, USA

Jonathan M. Nelson, U.S. Geological Survey, National Research Program, Denver, Colorado, USA

Gary Parker, St. Anthony Falls Laboratory University of Minnesota, Minneapolis, USA

Carlos Pirmez, Exxon Production Research Company Houston, Texas USA

Andrea Rinaldo, Dipartimento di Ingegneria Idraulica, Marittima e Geotecnica, Università di Padova, via Loredan 20, 35131 Padova, ITALY

Mark W. Schmeeckle, Department of Geology, Florida State University, Tallahassee, Florida, USA

Giovanni Seminara, Dipartimento di Ingegneria Ambientale, Università di Genova, Via Montallegro 1, 16145 Genova, ITALY

Ronald L. Shreve, Department of Earth and Space Sciences, UCLA, Los Angeles, California, USA

Tamotsu Takahashi, Disaster Prevention Research Institute, Kyoto University Gokasho, Uji, Kyoto 611-0011, JAPAN

1 Perspectives in Morphodynamics

Giovanni Seminara and Paolo Blondeaux

Dipartimento di Ingegneria Ambientale, Università di Genova, Via Montallegro 1 - 16145 Genova, Italy

1.1 Abstract

We introduce the reader to the subjects of the review papers by expressing some personal views on the paths that research on morphodynamics has recently undertaken and on the further developments which can be envisaged for the near future.

1.2 Morphodynamics: a new discipline?

Morphodynamics investigates natural patterns observed in the sedimentary environments of the Earth, which invariably arise from the mechanical interaction of sediments with the motion of natural fluids: as such, morphodynamics can hardly be considered as a new discipline. In fact, on one hand the latter definition identifies this subject as part of *geomorphology* and *geophysics*, on the other hand the interest of *hydraulic engineering* in the practical implications of morphodynamics is so old-dated that one can hardly trace its origin. Ample evidence of such interest is to be found in the activities of the International Association for Hydraulic Research (IAHR) since it was founded in 1935.

However, it is hardly disputable that in the last three decades morphodynamics has developed at a much faster rate than in the past: starting from the status of an essentially descriptive, empirically based discipline, it has progressively moved towards a more mature stage, attempting to become *predictive* and *quantitative*, i.e. to understand *why, when*, and *how much.*

The above progress, which has been due to the simultaneous efforts of various scientific communities, has seen a real discontinuity around the sixties when the first sound mechanical attempts were proposed to understand the mechanics of sediment transport and the nature of a variety of morphodynamical features of rivers and coasts (Kennedy [22]; Bagnold [7]; Callander [9]; Smith [38]; Engelund [15]). It is not surprising that, about the same time, date the first fundamental discoveries on the coherent structure of wall turbulence (Kline et al. [24]), whose relevance to the process of sediment transport are now being fully appreciated (see the contribution of Nelson et al. in this Volume). A number of researchers were involved in the above process but two of them, R. A. Bagnold and F. Engelund, stand over all the others for the vast influence that their original thinking exerted on the subsequent development of the subject. Indeed, though some of their cute ideas have not survived the challenge posed by detailed experimental observations and further theoretical developments (e.g. Seminara et al. [36]), however discoveries like *dispersive stresses* (Bagnold [6]), i.e. the mechanism able to support highly concentrated suspensions, or *bottom instability*, seen as leit motiv for the interpretation of fluvial and coastal patterns (Bagnold [8]; Engelund [15]; Engelund & Fredsøe [17]), will hardly lose their status of cornerstone contributions.

The above progress was also strictly related to the nearly simultaneous (though somewhat anticipated) development of some key ideas in the adjacent field of *geomorphology*. In particular, the cute observations of large scale fluvial forms independently performed in the late fifties by L. Leopold and his coworkers in U.S.A. (Leopold & Wolman [27]) and by R. Kinoshita in Japan (Kinoshita [23]) became the source of inspiration for a whole stream of researchers, which only in recent years has led to the successful development of models of meandering and braided rivers.

Though the hydrodynamics of the coastal region has distinct features (e.g. wave propagation, wave breaking, wave-current interactions) which make its study much richer than that of the quasi-steady phenomena typical of fluvial hydrodynamics, however viewing *river* and *coastal* morphodynamics as parts of a single discipline turns out to be a useful exercise. We have emphasized such viewpoint since the Euromech meeting 215 which we organized in S. Margherita Ligure on a similar subject roughly fourteen years ago. There is indeed ground to pursue analogies among mechanisms and patterns encountered in different subaerial or subaqueous environments. The mechanics of sediment mixtures in subaerial debris flows and mudflows (see the review paper by Takahashi in this volume) has analogies with the corresponding process (submarine debris flows) occurring in deep submarine environments, and has much in common with the so called sheet flow mode of sediment motion observed in the coastal environment. Small and large scale bedforms (dunes, bars, etc.) are observed in fluvial as well as in tidal channels. Turbidity currents generated in the deep submarine environments do not differ significantly from those occurring when a river flows into a reservoir or a lake. Meandering is an ubiquitous feature not only of subaerial (fluvial) and subaqueous (deep submarine) cohesionless environments (see the paper by Parker et al. in this volume) but also of partially cohesive environments like estuaries and lagoons.

The latter observations motivate the idea that gathering the different scientific communities involved in research on different aspects of our discipline may be a fertile approach leading to cross-fertilization internal to the subject.

1.3 Cross-Fertilization: a Chance or a Danger?

More recently the interest for morphodynamics has touched new scientific communities, namely those of *applied mathematicians and physicists*. The rationale behind such, perhaps unexpected, growth of interest lies in the complex nonlinear nature of morphodynamical systems which displays itself through the development of highly complex natural patterns providing an exciting territory for the exploration of classical and modern ideas on the behaviour of dynamical systems (Doelman & Van Harten [14]). The complexity of some of these patterns has also stimulated attempts to interpret their behaviour on the basis of new paradigms recently proposed as keys to understand 'how nature works' (see the paper by Rinaldo in this volume).

While there is no doubt that the use of the powerful tools offered by applied mathematics (perturbation methods and theory of dynamical systems above all) has allowed significant progress in the understanding of pattern formation, however it is fair to say that the danger exists of seeing an *uncontrolled* and *independent growth* of our interdisciplinary subject in different scientific contexts. Signs of the latter phenomenon are already present in the literature on morphodynamics which is increasingly widespread among a variety of journals with relatively little interaction between the respective audiences. Indeed, objective difficulties arise from the use of different, sometimes quite sophisticated languages, however we need recognize that the growth of fairly separated and selfmotivated disciplines does not serve the interests of science. It is not uncommon, though it is definitely undesirable, that the typical reader (or author) of practically oriented journals is unaware of the work published on similar subjects in more fundamentally oriented journals. Even more undesirable is the publication of papers where the chance to apply a known mathematical technique to a new problem prevails on the main scope of our work, which should consist of understanding new mechanisms and providing predicting tools for engineering purposes. Examples of such practice are unfortunately increasingly found in important, fundamentally oriented, journals where cellular models of complex systems are sometimes formulated based on a strikingly oversimplified picture of the physics of the system: the paradox of physicists with very little physics!

These are serious problems which have no simple solution and motivate our approach aimed at enhancing the interaction among all scientific groups involved in research on morphodynamics. It seems that such approach has been found fruitful as our proposal to have Symposia on River, Coastal and Estuarine Morphodynamics regularly organized every two years under the umbrella of IAHR has been enthusiastically accepted. RCEM 2 will be held in Nobihiro (Sapporo, Japan) on September 10-14, 2001.

1.4 New Trends of Research in Morphodynamics

What are the promising avenues of research in our field for the first decades of the new millennium? Generally speaking, we share De Vriend's opinion (this volume): just like most fields of science and engineering, research in morphodynamics has several legs, *field observations, laboratory experiments, numerical experiments and fundamental theoretical research.* In this respect we cannot agree about the statement sometimes made that fundamental theoretical research (in particular weakly nonlinear stability theories) should no longer be pursued and be replaced by numerical experiments, on the ground that nature is usually strongly nonlinear, hence weakly nonlinear approaches would have little practical relevance! The recent paper by Coleman & Fenton [12] rivisiting the classical potential flow approach of Kennedy [22] to explain the mechanism of dune formation has shown, if necessary, the need for a firm theoretical foundation of morphodynamic research. We have learnt a lot from Engelund's *linear* stability theories and the richness of the so called Landau and Ginzburg-Landau amplitude equations, which govern the weakly nonlinear development of unstable perturbations, suggest that a lot can be learned from *weakly nonlinear* morphodynamic analyses, just as it has occurred in several other branches of Fluid Mechanics. From the practical viewpoint, the least weakly nonlinear theories can do is to guide numerical experiments suggesting the existence of (possibly multiple) equilibrium solutions or, sometimes, warning against the possibility of chaotic responses. Here is an example where cross fertilization has definitely and will probably still play a very useful role.

A research topic, spanning all the areas of morphodynamics, which still awaits to be fully explored, is the construction of models of the *mechanics of sediment transport* based on a more detailed understanding of the hydrodynamics of particle-fluid interactions. Such need applies first of all to the mechanics of *particle saltation* occurring at fairly low Shields stress: models presently available are unable to interpret modern experimental observations of the effect of near wall coherent structures and still have to rely on Bagnold's assumption (Ashida & Michiue [2]; Engelund & Fredsøe [16]; Wiberg & Smith [41]), whose validity has been questioned several times and has been shown to be impossible when the bed is weakly sloping (Seminara et al. [36]). The need for significant developments extends to the *mechanics of suspension*, still commonly treated on the basis of the so called diffusion approximation, which definitely requires to be revisited in the light of the modern picture of near wall turbulent structures (Hogg et al. [20]). Such a need is even more dramatic in the field of highly concentrated suspensions of *poorly sorted granular mixtures*, a subject of fundamental interest for the understanding of debris flows, mudflows, sheet flows and turbidity currents. Here it is fair to say that, in spite of great efforts of several scientists, a sound model for the constitutive behaviour of such materials, able to combine the quasi static response displayed at low strain rates with the kinetic behaviour observed at high strain rates, with the generality required to describe the mechanics of two and three dimensional flows, is still lacking (Hutter et al. [21]).

The increasing role of technological developments in experimental as well as numerical techniques along with an improved theoretical insight will be needed to make significant progress in the quite complex field of research concerning *debris flows and mudflows*: indeed, in spite of the intense research activity devoted to this subject in the last few decades (Takahashi, this volume), the development of predictive models of the formation, propagation and stoppage of debris flows and mudflows able to account for the two phase nature of the flowing mixture with its complex rheological behaviour and the associated phenomena of particle sorting, channel excavation, levee formations will still require significant attention of the scientific community in the near future.

Fluvial meandering patterns have been thoroughly analyzed in the fluvial context (Ikeda & Parker, eds., 1989). Though new features of the meandering process have recently been disclosed (Zolezzi & Seminara [42]; Seminara et al. [35]) while other aspects (like meander formation in purely erosive environments, near bank processes and the occurrence of chute cutoff) are as yet not fully understood, it seems unlikely that the subject of river meandering will exhibit an overwhelming development in the near future. On the contrary, understanding the mechanics of braiding is the great challenge of river morphodynamics, which has received recent attention through attempts to analyze some of the fundamental processes operating in the development of braiding (Seminara et al. [36]) or by means of numerical models of the evolution of the whole network (Murray & Paola [30]; Engrgrob & Tjerry [18]; Mc Ardell and Faeh [29]).

Tidal and estuarine morphodynamics is also vastly unexplored and calls for the development of a better understanding of the dynamics of partially cohesive environments and of the effects of vegetation on morphodynamics. In fact, the recent literature shows that the analysis of several basic processes which have been deeply investigated in the fluvial context, like the equilibrium profile of cohesionless channels, the equilibrium cross section of cohesionless channels, the formation of small and large scale bedforms, meander formation and the evolution of fluvial networks, have not been extended to the tidal case so far, except for some, not yet conclusive, exceptions (Schuttelaars & de Swart [31]; Seminara & Tubino [37]; Lanzoni & Seminara [26]). Knowledge in this field is then still mainly based on observations of geomorphologists and on the practical expertise that hydraulic engineers have developed in the management of estuaries and lagoons.

Coastal morphodynamics has been the subject of an intense research activity in the last few decades, in particular in Europe the EU has funded several research projects which have involved most of the European hydraulic community.

This activity has led to substantial improvements of both our knowledge of the physical processes controlling the mechanics of coastal forms (Blondeaux [5]) and our ability to predict the time development of the morphology of the coastal region using complex models (see Fredsoe & Deigaard, present volume).

On the theoretical side, the linear stability analyses of basic coastal morphologies, aimed at investigating the mechanisms of appearance of rhythmic patterns of the seabed and/or the shoreline, have been slowly replaced by attempts to incorporate nonlinear effects. In particular the interaction of different components of the perturbation has been investigated to explain the formation of complex topographies (Vittori & Blondeaux [39]; Schuttelaars & De Swart [32]) and the nonlinear coupling between hydrodynamic and morphodynamic modes has been considered to understand the formation of particular morphological patterns (Vittori et al. [40]; Coco et al. [11]). Further research is required to explain the simultaneous appearance of bedforms of different scales through a self-exciting mechanism (see the recent contribution by Komarova & Newell [25]). Attempts have also been made to predict the equilibrium configuration of the bottom by means of fully nonlinear models, which are usually fairly complex and require advanced numerical techniques (Andersen [1]; Calvete et al. [10]). However this line of research is still at its infancy and significant progress will be needed in order to attain reliable results.

The stochastic forcing (storms) driving sediment motion in the coastal region along with the intrinsic unpredictable character of some morphodynamic systems, which are characterized by a highly nonlinear dynamics, have posed a new challange to hydraulic engineers, namely the long term prediction of the morphological changes experienced by the coastal region (see De Vriend present volume). Remote sensing techniques, like the video monitoring system introduced by Lippmann & Holman [28] (see also Holman, present volume), are valuable tools to obtain high resolution information of coastal morphology required to test the available models.

Further fundamental research on turbulence and sediment dynamics is also needed. In fact, even though large scale morphodynamic models usually need to quantify only the sediment transport rate averaged over the wave cycle, a detailed knowledge of the intra-wave processes is required to obtain reliable estimates. Production of turbulence which takes place close to the sea bed during the decelerating phases of the wave cycle (Hino et al. [19]) and the related peaks of sediment concentration observed at flow reversal (Davies et al. [13]) are known phenomena which are still to be accurately modelled. These peculiar aspects of sediment transport driven by unsteady flows add further complexity to the problem of understanding the interaction between fluid and sediment motion.

Deep submarine morphodynamics (see the contribution of Parker in this volume) is still at its infancy, though the interest of oil companies in the structure of submarine alluvial fans and the increasing practice to choose paths for submerged pipelines which border the edge of the continental shelf will likely enhance research in this field in the near future.

1.5 Reductionist Versus Holistic Approaches: a False Problem

Let us conclude by mentioning a recently agitated question, regarding what is seen by someone as a conflict between two different ways to approach research in our field: a so called *reductionist* approach, based on the idea that understanding the behaviour of a complex system requires that the fundamental mechanisms controlling the dynamics of its parts must be firstly at least qualitatively understood, versus a so called *holistic* approach, which assumes that the complex nature of the system can be investigated by tools which describe its overall behaviour and are not based on a previous understanding of the physical processes operating at the small scale level. An echo of the reductionist-holistic debate is also felt in the contribution of De Vriend in this volume.

A splendid example of the former approach is the work on suspension rheology of a giant of science, G.K. Batchelor [4], where the rheological (macroscopic) behaviour of fairly dilute suspensions was derived from a detailed understanding of the hydrodynamics of particle motion and of particle-particle interactions. A popular example of the latter approach is the so called *theory of selforganized criticality* (Bak [3]).

Our feeling is that in judging any approach we must rely on epistemology: the word 'scientific' in Popper's sense, implies that any statement must be amenable to some kind of 'falsification' procedure. In this respect any 'scientific' contribution should be welcomed by the scientific community even if it only contributes to enlightening a very narrow aspect of a complex phenomenon. We do not see any conflict between reductionist and holistic approaches as long as they are truly 'scientific'. However, the weakness of some of the recent so called 'holistic' approaches lies in their tendency to progressively abandon the need to provide a sound physical basis for the formulations proposed, a tendency which should be discouraged as it locates such research in the realm of metaphysics rather than in the field of science. Admittedly, a related danger exists in the area of reductionism, namely that of analyzing detailed mechanisms in the context of too idealized models of reality: such research has no significant impact and keeps its interest confined within the realm of the academic world. It will be our task to maintain an equilibrium between the latter two extreme attitudes.

References

1. Andersen, K.H. (1999) The dynamics of ripples beneath surface waves and topics in shell models of turbulence. Ph.D. Thesis, Univ. Copenhagen.
2. Ashida, T.R., Michiue, M. (1972) Study on hydraulic resistance and bedload transport rate in alluvial streams. Transactions Japan Society of Civil Engineering, **206**, 59-69.

3. Bak, P. (1996) How nature works: the science of self-organized criticality. New York. Copernicus-Springer.
4. Batchelor, G.K. (1970) The stress system in a suspension of force-free particles. J. Fluid Mech., **41**, 545-570.
5. Blondeaux, P. (2001) Mechanics of coastal forms, Ann. Rev. Fluid Mech., **33**, 339-369.
6. Bagnold, R.A. (1954) Experiments on a gravity free dispersion of large solid spheres in a Newtonian fluid under shear. Proc. Roy. Soc. Lond. Ser. A, **225**, 49-63.
7. Bagnold, R.A. (1966) An approach to the sediment transport problem from general physics. Geological Survey Professional Paper, 422-I, Washington DC.
8. Bagnold, R.A. (1946) Motion of waves in shallow water. Interaction of waves and sand bottoms. Proc. Roy Soc. London, A.,**187**, 1-15.
9. Callander, R.A. (1968) Instability and rivers meanders. Ph.D. Thesis, Univ. Auckland.
10. Calvete, D., Falques, A., de Swart, H.E., Dodd, N. (1999) Nonlinear modelling of shoreface-connected sand ridges. Coastal Sediments, A.S.C.E., June 20-24, 1123-1138.
11. Coco, G., Huntley D.A., O'Hare, T.J. (2000) Investigation of a self-organization model for beach cusp formation and development. J. Geophys. Res., **105**, C9, 21991-22002.
12. Coleman, S.E., Fenton, J.D. (2000) Potential-flow instability theory and alluvial stream bed forms. J. Fluid Mech., **418**, 101-117.
13. Davies, A.G., Ribberink, J.S., Temperville, A., Zyserman, J.A. (1997) Comparison between sediment transport models and observations made in wave and current flows above plane beds. Coastal Engineering, **31**, 163-198.
14. Doelman, A., Van Harten, A. (1995) Nonlinear dynamics and pattern formation in the natural environment. Pitman Research Notes in Mathematics Series, Longman.
15. Engelund, F. (1970) Instability of erodible beds. J. Fluid Mech., **42**, 225-244.
16. Engelund, F., Fredsøe, J. (1976) A sediment transport model for straight alluvial channels. Nordic Hydrology, **7**, 293-306.
17. Engelund, F., Fredsøe, J. (1982) Sediment ripples and dunes, Ann. Rev. Fluid Mech., **14**, 13-37.
18. Enggrob, H.G., Tjerry, S. (1999) Simulation of morphological characteristics of a braided river. Proc. of IAHR Symposium on River, Coastal and Estuarine Morphodynamics, Genova, 6-10 settembre, vol. I, 585-594.
19. Hino, M., Sawamoto, M., Takasu, S. (1976) Experiments on transition to turbulence in an oscillatory pipe flow. J. Fluid Mech., **75**, 193-207.
20. Hogg, A.J., Huppert, H.E., Soulsby, R.L. (1993) The dynamics of particle-laden fluid elements. In: Euromech 310 Sediment Transport mechanisms in in coastal environments and rivers, Le Havre.
21. Hutter, K., Svendsen, B., Rickenmann, D. (1996) Debris flow modeling: A review. Continuum Mech. Thermodyn., **8**, 1-35.
22. Kennedy, J.F. (1963) The mechanics of dunes and antidunes in erodible-bed channels, J. Fluid Mech., **16**, 521-544.
23. Kinoshita, R. (1961) An investigation of channel deformation in the Ishikari River. Technical Report, Natural Resources Division, Ministry of Science and Technology of Japan.

24. Kline, S.J., Reynolds, W.C., Schraub, F.A., Rundstadler, P.W. (1967) The structure of turbulent boundary layers, J. Fluid Mech., **30**, 741-773.
25. Komarova, N.L., Newell A.C. (2000) Nonlinear dynamics of sand banks and sand waves, J. Fluid Mech., **415**, 285-321.
26. Lanzoni, S., Seminara, G. (2001) Long term evolution and morphodynamic equilibrium of tidal channels. Submitted for publication on J. Geoph. Res.
27. Leopold, L.B., Wolman, M.G. (1957) River channel pattern: braided, meandering and straight, U.S. Geol. Survey, Prof. Paper 282-B.
28. Lippmann, T., Holman, R. (1989) Quantification of sand bar morphology: a video technique based on wave dissipation. J. Geophys. Res., 94, 995-1011.
29. Mc Ardell, B.W., Faeh, R. (2000) A computational investigation of river braiding. Proc. Gravel-bed Rivers 2000, Christchurch, New Zealand.
30. Murray, A.B., Paola, C. (1994) A cellular model of braided rivers. Nature, **371**, 54-57.
31. Schuttelaars, H.M., de Swart, H.E. (1996) An idealised morphodynamic model for a tidal embayment. Eur. J. Mech. B/Fluids, **15**, 55-80.
32. Schuttelaars, H.M., de Swart, H.E. (1999) Formation of channels and shoals in a short tidal ambayments. J. Fluid Mech. **386**, 15-42
33. Seminara, G., Blondeaux, P. (eds.) (1999) Proc. of IAHR Symposium on River, Coastal and Estuarine Morphodynamics, Genova, 2 volumes.
34. Seminara, G., Solari, L. and Parker, G. (2000) Trasporto solido a bassi Shields e non validità dell'ipotesi di Bagnold. XXVII Convegno di Idraulica e Costruzioni Idrauliche, vol. 1, 349-356, Genova (in Italian).
35. Seminara, G., Zolezzi, G., Tubino, M., Zardi, D. (2001) Upstream and downstream influence in river meandering. Part two: planimetric development. To appear in the J. Fluid Mechanics.
36. Seminara, G., Tubino, M., Paola, C. (2000) The morphodynamics of braided rivers: experimental and theoretical results on unit processes. Proc. Gravel-bed Rivers 2000, Christchurch, New Zealand.
37. Seminara, G., Tubino, M. (2001) Sand bars in tidal channels. Part one: tidal free bars. To appear in the J. Fluid Mechanics.
38. Smith, J.D. (1970) Stability of a sand wave subjected to a shear flow of low Froude number. J. Geoph. Res., **75**, 5928-40.
39. Vittori, G., Blondeaux, P. (1992) Sand ripples under sea waves. Part 3. Brick-pattern ripple formation. J. Fluid Mech., **239**, 23-45
40. Vittori, G., De Swart, H.E., Blondeaux, P. (1999) Crescentic bedforms in the nearshore region. J. Fluid Mech., **381**, 271-303
41. Wiberg P.L., Smith J.D. (1985) A theoretical model of saltating grains in water. J. Geoph. Res., **90**, 7341-7354.
42. Zolezzi, G., Seminara, G. (2001) Upstream and downstream influence in river meandering. Part one: upstream overdeepening. To appear in the J. Fluid Mechanics.

2 Sediment Entrainment and Transport in Complex Flows

Jonathan M. Nelson[1], Mark W. Schmeeckle[2], Ronald L. Shreve[3], Stephen R. McLean[4]

[1] U.S. Geological Survey, National Research Program, Denver, Colorado USA
[2] Department of Geology, Florida State University, Tallahassee, Florida, USA
[3] Department of Earth and Space Sciences, UCLA, Los Angeles, California, USA
[4] Department of Ocean Engineering, UCSB, Santa Barbara, California, USA

2.1 Abstract

Predicting the entrainment and transport rates of sediment grains making up an erodible bed underlying an arbitrary flow field requires a mechanistic understanding of the coupling between the flow and the forces on sediment grains. To help develop such an understanding, a suite of flow and sediment-transport experiments are described; these may be loosely divided into two categories. First, measurements of near-bed flow structure and sediment motion in a variety of spatially or temporally accelerating flows are used to show the manner in which changes in flow structure can impact sediment entrainment and transport. Second, direct high-frequency measurements of lift and drag on sediment particles in various turbulent flows are used to make a more direct connection between near-bed flow structure and sediment dynamics. Taken together, these experiments show how even changes in turbulence structure due to spatial and/or temporal accelerations can have a significant effect on the sediment-transport field. Finally, a method is briefly outlined for predicting sediment motion under arbitrary flows using either measured near-bed velocity time series or flow information predicted from direct numerical simulations or large-eddy simulations.

2.2 Introduction

The shape and sedimentological content of landforms made up of wind or water borne sediments is ultimately the result of the complex interactions between flow, sediment transport and bed morphology. The goal of scientists and engineers working on various aspects of this flow-bed-sediment coupling is to develop an understanding of the governing physics that is adequate for developing predictive models for certain features. For example, fluvial geomorphologists and river engineers want to predict the shape of river channels, bars and bedforms given flow and sediment supply and valley slope. Over the past decade or so, impressive advances have been made in our abilities to make accurate predictions of the morphology and behavior of certain geomorphic features (notably riverine and coastal barforms), primarily through the development and application of coupled computational models for flow, sediment transport, and bed evolution. However, there are several situations for which these coupled models perform poorly, even for cases where the flow field solution is accurate, suggesting that the sediment-transport algorithms are at fault.

Currently, the sediment-transport algorithms that are used in computational bed evolution models typically incorporate (1) an empirical bed-load relation that predicts local sediment flux as a function of local particle sizes and some local flow parameter (typically bed stress), and (2) a suspended load advection-diffusion calculation that yields concentration by size class and local erosion or deposition using a lower boundary condition relating near-bed concentration or upward sediment flux to flow parameters (bed stress or near-bed turbulence intensity). However, it is important to note that these algorithms were developed and calibrated in simple steady, uniform flows, where all the near-bed turbulence statistics scale strongly with the boundary shear stress. This scaling does not generally hold in flows with temporal and/or spatial acceleration, and there is no reason to expect that relations developed near the bed of very simple boundary layers should be accurate in more complex flows.

To investigate the possibility that the error in sediment flux prediction is associated with the decoupling of boundary shear stress and other turbulence statistics in unsteady or nonuniform flows, several experiments were carried out. These experiments treat several situations that are important in natural sediment-transporting flows, including flow behind a backward step, flow over bedforms, and flow in an oscillatory boundary layer. Laser- and acoustic-Doppler velocimetry measurements from these experiments show that spatial or temporal acceleration has a significant effect on near-bed flow characteristics, and that flows with identical values of boundary shear stress can have significantly different turbulence structure. Furthermore, digitizing sediment motion from high-speed motion picture photography of the bed clearly shows that flows with the same boundary shear stress (but different spatial or temporal acceleration) can have local sediment flux values that differ by as much as an order of magnitude. These results suggest that at least some portion of the variability found in sediment-transport measurements is not associated with some "unknowable" aspect of the problem, but is a direct result of our own poor understanding and the associated parameterization. The experiments also show that local sediment flux is highly correlated with certain aspects of the near-bed flow but, as one might expect, these aspects are associated with relatively rare events in the flow distribution, which are poorly captured in mean quantities such as bottom stress, and are greatly affected by accelerations.

To develop a clearer understanding of the processes governing sediment entrainment and motion, and noting that the forces of lift and drag ultimately move sediment particles, we present direct measurements of horizontal and vertical force on a variety of stationary particles in several uniform and nonuniform flows. The measurements were made using a newly developed transducer, which measures forces on particles at turbulence resolving frequencies (up to 200 Hz). The measurements corroborate our more indirect determination of the effects of acceleration. They show that the distributions of lift and drag on particles on a sediment bed are strongly dependent on both the magnitude and frequency content of the near-bed velocity distributions, which are in turn sensitive to rates of local acceleration. More importantly, the information from the force transducer allows construction of models which predict particle motion from a direct computation of forces using a predicted or measured velocity time series, i.e., without using some averaged surrogate for the flow information, such as bed stress. This mechanistic approach yields flux predictions that are in good agreement with empirical models and data from steady, uniform flows, but it also allows appropriate computation of sediment transport in complex flows. The model does not require the arbitrary specification of a Bagnold or Owen type boundary condition (requiring the Reynolds stress to be the critical shear stress at the bed due to momentum extraction from the flow by moving particles), because it depends on the statistics of the velocity distributions near the bed; most entrainment and transport results from relatively rare events in the flow, not mean conditions.

In summary, the experimental results presented in this paper along with computational approaches support the hypothesis that boundary shear stress alone is not an ideal predictor of sediment flux in all but the simplest flows. However, it is possible to develop more accurate, general relations between near-bed flow and sediment motion that proceed directly from simple force balances. These relations allow the prediction of sediment motion, and the associated bed evolution, even for complex nonuniform and unsteady flows. Although these techniques are in their infancy, they hold great promise for expanding the range of flow and sediment-transport situations for which we can make accurate predictions of the evolution of bed morphology. This is especially true in natural flows, which almost always include strong spatial and temporal accelerations, and which are, after all, the real domain of interest.

2.3 Flow and Sediment Transport Experiments

In this section, several types of experimental data are used to examine the coupling between flow structure and sediment motion. The experimental results are briefly discussed in historical order, and it should be clear that the each experiment was motivated by the results of the previous one, as one might expect. Most of these results are described in depth elsewhere, and will only be used here to provide a framework for considering the role of turbulence in sediment entrainment and transport.

2.3.1 Flow and sediment transport over bedforms

Two and three-dimensional bedforms are the most common erodible bed feature in natural flows. However, despite their importance, there is still no general theoretical approach for predicting their shapes or sizes in arbitrary flows. As a result, practical methods for determining bedform morphology rely heavily on empirical data. While this is adequate for simple flows, it is less than ideal for predicting bedform shapes and sizes in unusual or strongly varying flows. Predicting the behavior of bedforms from first principles in an arbitrary flow requires that we understand and characterize the complex coupling between the flow field, the sediment-transport field, and the wave-like bed geometry, which is a difficult task. Even though bedforms and their effects on the flow are often considered to be weak perturbations about a simple steady, uniform flow, careful observation shows that these features have a profound influence on both the mean flow and turbulence fields.

The first step in understanding the formation and evolution of bedforms is to understand the flow over these features. With this in mind, many investigators have used a variety of techniques to measure detailed velocity fields over bedforms. Using such measurements, Raudkivi [15] pointed out that the spatial evolution of the turbulence field over a typical bedform appeared to play a central role in the pattern of sediment entrainment and deposition over such a feature. The spatial evolution of the turbulence field over most bedforms, such as those studied by Raudkivi, is dominated by the processes of flow separation and reattachment, along with wake formation and shear-layer relaxation. Thus, even though bedforms are often considered to be roughness elements on the boundary of a simple uniform flow (at least in a spatially averaged sense), their local effects dominate the structure of the turbulence field; Raudkivi inferred that this would have a major effect on the sediment-transport field.

Over the ensuing 35 years, advances in instrumentation and computational techniques have provide a somewhat clearer view of the mechanics of flow over bedforms and other similar erodible bed features. These advancements have served to emphasize the role of wake turbulence in the entrainment and transport of sediment, and to show why that structure is important. In Figure 2.1, laser-Doppler measurements of mean velocity over a simple two-dimensional bedform are shown (from Nelson and Smith [11]). The signatures of flow separation and reattachment are clear, as is the presence of a diffusing momentum defect on the stoss (upstream side) of the bedform. The momentum defect associated with the wake of the upstream bedform overlies a growing internal boundary layer. The classical view of this situation is that the skin-friction boundary shear stress at the base of the internal boundary layer should characterize the entrainment of sediment from the bed, and

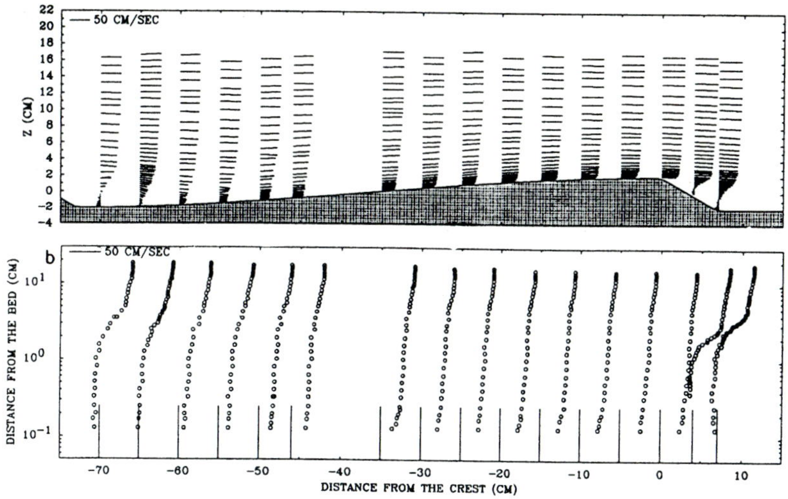

Fig. 2.1. Laser-Doppler velocimeter measurements of the mean flow field over a two-dimensional bedform. Vector velocities on a linear scale are shown in (a) and streamwise velocities are shown on a logarithmic scale in (b).

that this quantity is an accurate predictor of bedload sediment flux over the bedform stoss side. However, these ideas are predicated on results found for steady, uniform flows, where the near-bed turbulence statistics scale with the boundary shear stress. Looking in more detail at the turbulence structure (Figure 2.2) shows that the Reynolds stress field over much of the bedform is dominated by turbulence generated in the free-shear layer downstream of the separation point at the bedform crest, as pointed out in several experimental investigations (Nelson and Smith [11], Nelson et al. [12], McLean et al. [8], Bennett and Best [2]). Although this flow is often considered to be analogous to a boundary layer, many aspects of the flow are actually more wakelike. If one considers the structure of the wake flow, the bed is analogous to the centerline of a two-dimensional wake. In this region of a wake, the shear stress (momentum flux) is zero, but the turbulence intensity is maximum. In other words, if there were not a no-slip boundary condition at the bed, the correlation coefficient would be zero, as it is at a wake centerline. Keeping this in mind, but also noting the requirement for a thin internal boundary layer to enforce the bottom boundary condition, one concludes that the near-bed flow can be thought of as an internal boundary layer forced by relatively large-scale structures which are ineffective for transporting momentum in the vertical direction. This simple reasoning leads to exactly what one observes in making measurement very near the bed; there is a very well-formed internal boundary layer with typical vertical structure (near-logarithmic), but the

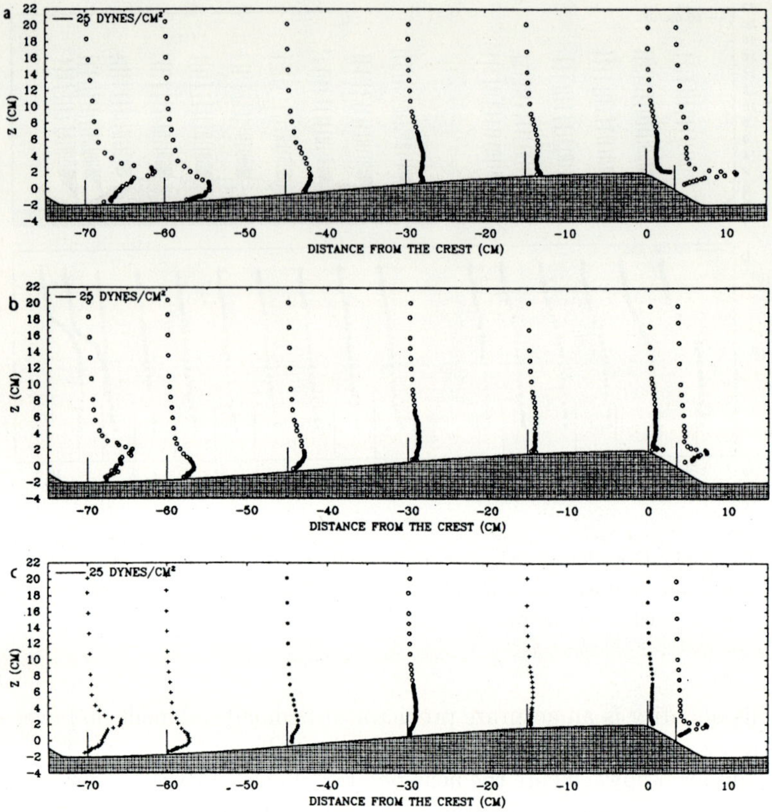

Fig. 2.2. Measured components of Reynolds stress over a two-dimensional bedform. Streamwise, vertical, and shear components are shown in (a), (b), and (c), respectively.

turbulence intensities in that layer are very high relative to the shear stress, at least compared to a conventional uniform boundary layer. In Figure 2.3, the correlation coefficient is plotted from the measurements shown in Figure 2.2. Over the entire stoss of the bedform, the values are significantly lower than that found in well-developed boundary layers (typically 0.4), reflecting the role of overlying wake turbulence on the boundary layer structure. Values computed from the near-bed measurements only approach this value at the crest of the bedform. As lift and drag are nonlinear with streamwise velocity, this additional variability in u and w for a given value of bed stress means one would expect higher sediment-transport rates in this situation relative to another flow with the same boundary shear stress but higher correlation

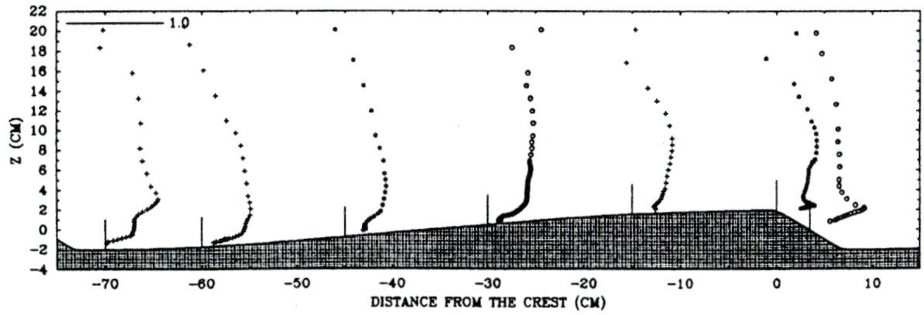

Fig. 2.3. Correlation coefficients for flow over a bedform using data shown in Figure 2.2.

coefficients near the bed. This is a clear example showing how spatial variability can potentially produce different rates of sediment transport for the same values of boundary shear stress.

2.3.2 Flow and sediment transport downstream of a backward step

In order to better understand the role of flow separation and, more specifically, the impact of the wake downstream of a separation point on the internal boundary layer structure, measurements on both flow and sediment transport were carried out downstream of a backward step. The experiment is described in detail in Nelson et al. [13] and is shown schematically in Figure 2.4. Simultaneous near-bed flow and bedload transport measurements were obtained for various degrees of flow nonuniformity and kinds of turbulence structure by making measurements using a plane bed downstream of a backward (i.e., downstream-facing) step about 4-cm high in a flow 20-cm deep, as well as with the step removed. The bed was composed of 0.9-mm sand screeded flat and subjected to transport before each run. Laser-doppler velocimetry (LDV) and high-speed motion-picture photography were used to measure simultaneously and noninvasively both fluid flow and sediment movement at the frequencies associated with turbulence. The LDV measurements were made 5 mm above the bed at average rates of about 250 Hz. Each measurement consisted of the time of measurement and the streamwise (u) and bed-normal (w) components of the velocity. Grain motion was recorded using a 16-mm camera running at 240 frames per second. The camera and LDV clocks were synchronized electronically to within 1 ms. An area of the bed about 5 cm square was filmed from above; and the fluid velocities and sediment fluxes were measured at the center of the film images. The sediment fluxes were measured by counting the grains that crossed a 10-mm cross-stream line during successive 0.1-s intervals; and the corresponding fluid velocities were computed by averaging the unequally spaced LDV measurements. The resultant 10-Hz time series comprise the basic data analyzed.

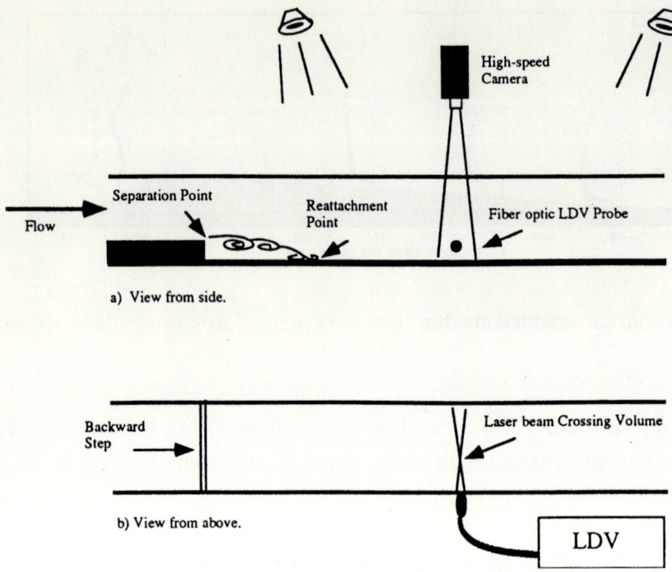

Fig. 2.4. Schematic of laboratory setup for measuring near-bed flow and sediment transport downstream of a backward step. The view from the side and from above are shown in (a) and (b), respectively.

Basic data from the location ten step heights from the step (Run 1) and the case with no step (Run 7) were significantly different in both turbulence structure and mode of sediment transport (Figure 2.5). The turbulence intensity was much greater and the sediment transport rate was more variable in Run 1. Furthermore, although near-bed velocity and average transport rate were greater in Run 7, the peak 0.1-s transport rates were higher in Run 1. The transport was strongly episodic, with short periods of high transport interspersed with long periods of relatively low or, commonly, zero transport, particularly in Run 1.

The extreme variability in transport can be partially attributed to the fact that the bed shear stress was only about twice the critical stress for significant motion. The differences between Runs 1 and 7, however, appear to be largely due to differences in turbulence structure. The numerous relatively low-frequency fluctuations in Run 1 were not present in Run 7. They arise from the vortex shedding and wake flapping associated with flow separation over a sharp break in bed elevation, as discussed by many authors (Bradshaw and Wong [3], Eaton and Johnstone [5], Simpson [21]) in general and by Nelson et al. [12] specifically in the context of sediment transport over bedforms. Joint frequency distributions of the 10-Hz streamwise and vertical velocity fluctuations for Runs 1 and 7 are shown in Figure 2.6. The quadrants in the diagrams correspond to the four kinds of turbulence events that

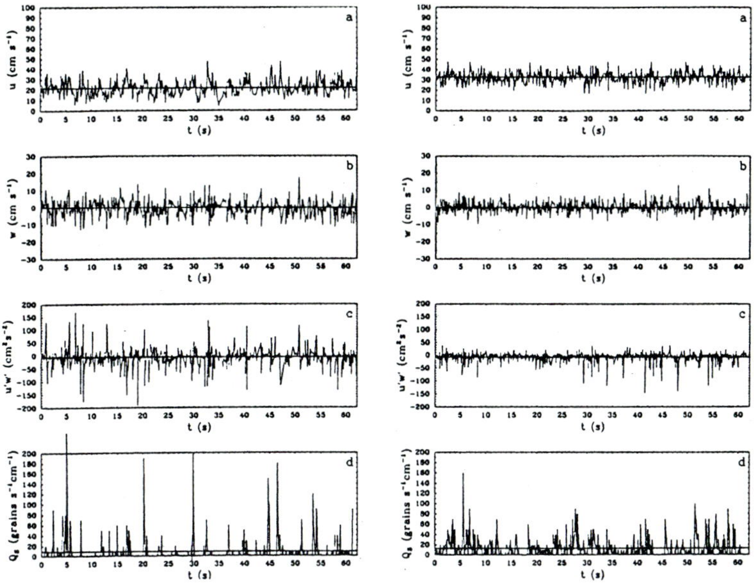

Fig. 2.5. Time series of data for Runs 1 (10 step heights from the step; left) and 7 (no step; right). Quantities plotted are (a) streamwise velocity, (b) vertical velocity, (c) upwards Reynolds momentum flux, and (d) sediment transport rate.

characterize the individual 10-Hz velocity measurements: bursts (u' less than 0, w' greater than 0), sweeps (u' greater than 0, w' less than 0), outward interactions (u' and w' greater than 0), and inward interactions (u' and w' less than 0). Bursts and sweeps contribute positively to the bed shear stress, whereas outward and inward interactions contribute negatively to it. Thus, because the local bed shear stress is positive, the long axes of the distributions are rotated so that more points are in the burst and sweep quadrants than in the interaction quadrants. The principal change in the flow caused by the step is the presence of a greater number of relatively high-magnitude velocity fluctuations uniformly distributed in the four quadrants, and hence ineffective for altering the mean bed shear stress, in accord with the observed lack of much spatial variation in the bed shear stress in the region of the runs.

Similar diagrams (not shown, see Nelson et al. [13]) in which the 10-Hz velocity measurements are weighted according to the corresponding sediment transport rate show that the transport rate and streamwise velocity are strongly positively correlated and that the majority of sediment is moved by sweeps and outward interactions, rather than by bursts and inward interactions. Thus, bursts and sweeps, or inward and outward interactions, transmitting exactly the same momentum fluxes move widely different amounts of sediment. On average the transport per 0.1 s during sweeps approximately equals that during outward interactions and greatly exceeds that during bursts and inward interactions. Sweeps move more sediment than outward interactions only because they are more common: individually, outward interactions are just as effective, even though they contribute negatively to the bed shear stress.

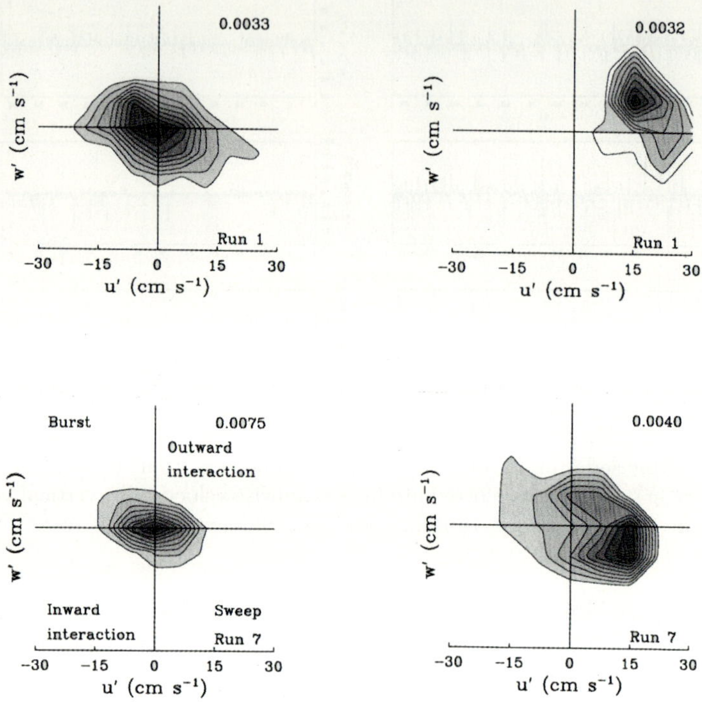

Fig. 2.6. Joint distributions of u' and w' at 10Hz (two left diagrams) and average transport rate for 0.1-s intervals (two right diagrams) for Runs 1 and 7. The distributions are normalized to peak values of one. For the left diagrams, multiplying the contoured values by the number in the upper right corner of the diagram yields a standard probability density diagram. For the diagrams on the right, integrating the contoured values over any area of the diagram and multiplying by the number in the upper right-hand corner yields the ratio of the sediment transport in that area of the diagram to the total transport.

Figure 2.6 also shows the average transport rate for 0.1-s counting intervals with specified fluctuations from the mean velocity for Runs 1 and 7. The diagrams indicate that the events most effective for transporting sediment during 0.1-s intervals are those having the faster streamwise velocity fluctuations, with high-magnitude outward interactions being most effective close to the step and sweeps being most effective farther downstream and in the absence of a step. This shift is primarily due to the decreasing magnitude of the outward interactions as the large, low-frequency eddies associated with flapping of the free shear layer and vortex shedding from the separation point dissipate. This is confirmed by the films, which show that moving sediment rose higher nearer the reattachment point.

As anticipated from the previous work on fixed bedforms by Nelson et al. [12] and McLean et al. [8], the measurements show a significant peak in bedload transport even though the bed shear stress changes little in the region of the experimental runs downstream of the reattachment point. This peak in transport was observed about 20 step heights downstream of the step, and the local flux at that point was about 2.5 times the flux measured with the step removed (Nelson et al. [13]). By delineating the turbulence events that move the most sediment, the measurements make clear that characteristics of the frequency structure of the turbulence, such as systematic variation in the duration of turbulence events, can significantly influence the spatial pattern of sediment transport. They also demonstrate that the spatial pattern of turbulence structure downstream of a reattachment point will produce a peak in bedload transport. Thus, any bed feature that causes separation in a flow transporting bedload will cause deposition of a bump downstream, which would in turn cause deposition of another bump, and ultimately the development of a whole chain of bedforms. Moreover, growth of the bedforms eventually would be limited, and stability achieved, by changes in the turbulence field caused by the increasing topographically induced acceleration (see Nelson et al. [13] for a more detailed discussion).

Careful experiments consistently show that separation plays an important role in the development of bedforms on initially flat beds (Raudkivi [14,15], Williams and Kemp [28]); the observations downstream of the step show how. More generally, they show how spatial nonuniformity can produce variations in turbulence structure not characterized by local bed shear stress and indicate that the resulting effects may play critical roles in the mechanics of erodible beds.

2.3.3 Flow and sediment transport in an oscillatory boundary layer

The above experiments show how spatial nonuniformity can change the relationship between local bed stress and local sediment flux, but they do not address similar issues in unsteady flows. To examine the potential effects of temporal accelerations in producing similar effects, an experiment was carried out to measure bedload transport and near-bed flow over a flat bed in an oscillatory flow duct. Sediment flux and velocity were determined at 100hz using synchronized high-speed cinematography and laser-Doppler velocimetry, respectively. We used 0.8 and 1.3 mm well-rounded particles, with Shields numbers of about 0.08 (based on maximum stress), corresponding to wave

the bed results in a very accurate predictor for sediment flux at all phases of the wave (Figure 2.12). To some degree, this figure demonstrates that sediment flux is predictable even in complex flows. In a more general sense, all the above experimental results suggest that there is a fairly tight coupling between transport and relatively large streamwise velocities near the bed. Given that mean lift and drag are strongly related to the streamwise velocity, this makes good sense, and it suggests a logical course of action to investigate the problem further.

2.4 Force Experiments

The measurements of near-bed flow and bedload sediment fluxes obtained using a combination of synchronous laser-Doppler velocimetry (LDV) and high-speed photography show that the coupling between flow and sediment entrainment and motion that appears to work well in steady, uniform boundary layers is much less accurate in nonuniform or unsteady flows. However, this does not mean there is not a well-defined relation between the near-bed flow and the transport. In fact, all the experiments suggest that there is a clear relationship, but that quantifying the relationship requires that we specifically consider how near-bed velocity fluctuations affect sediment particles. To make a deterministic relation that predicts transport or vertical fluxes of sediment from the bed, it is necessary to relate variability in the velocity to variability in the forces exerted on the sediment grains making up the bed.

With this in mind, the next step is to develop a better understanding of the coupling between near-bed flow dynamics and the forces exerted by the flow on sediment grains making up an erodible bed. In the past, there have been many investigations describing the coupling between the mean flow characteristics and mean forces on grains, but there have been few measurements of the temporal variability and frequency distributions of forces on grains in a turbulent flow. Apperley and Raudkivi [1] measured time series of horizontal and vertical force, and were able to determine the frequency response of the grain and the variability in the measured forces, both of which are important pieces of the overall problem of flow-sediment coupling. In our experiments, we extend this idea to include velocity measurements made with an LDV, which allows examination of the relation between flow and forces on grains in a near instantaneous sense.

To accomplish this, we interfaced a newly developed force transducer with an LDV to measure horizontal components of velocity and force at 200 hz. Data were collected in a variety of flow situations, including well-developed smooth and rough turbulent boundary layers, and interaction zones with both wake-like and boundary layer characteristics. We used several test particles, including spheres, cubes, and several different natural grains.

2.4.1 Experimental Procedure

The equipment used and experimental setup are shown in Figure 2.13. The force transducer and the interface between the transducer and the LDV were developed through a cooperative effort between Dr. William Bruner of Hylozoic Products and the U.S. Geological Survey. The device was specifically designed to resolve high frequency forces and was calibrated in two different ways. First, a simple calibration for force was determined using calibrated weights for each of the two dimensions that the transducer measures; each channel was found to be linear to a high degree of accuracy (R greater than 0.99). Second, a set of two orthogonal coils was used to generate time-varying forces on a magnetic test particle. Using a frequency generator to drive the coils, we measured the frequency response using both simple sinusoidal forcing at a variety of frequencies and by using square wave forcing. We determined the frequency response of the two channels up to 200 hz, and also carefully tested for any cross-coupling effects between the two channels (dimensions). The frequency response was essentially perfect up to around 30 hz, and was well-behaved above that level, with modest peaks (2 to 3 times nominal amplitude) at 60-70 hz and a slow decrease to zero at over 200 hz. The LDV sampling volume was typically located either immediately (3 particle radii or half-widths) upstream or immediately (2mm) above the particle, which was 1.9 cm in diameter. Reference velocities for drag calculations were corrected for the presence of the particle using potential flow theory, and the correction procedure was checked by measuring at a variety of distances upstream of the particle. This correction is necessary as measurements made upstream of the particle were generally made close enough for the measured velocity to be influenced by the particle; the correction to free-stream velocity was less than 5%. Measured force time series were filtered using a digital filter designed to correct for the frequency response of the transducer.

Figure 2.14a shows typical time series of the horizontal and vertical components of velocity and force. From these, we computed mean drag and lift, drag coefficients, distributions of drag and lift, correlation coefficients between the components of velocity and force, and other statistical and spectral information.

2.4.2 Mean flow coupling

As shown in Figure 2.14b, the drag coefficients we measured for spheres in a turbulent flow near both hydraulically smooth and hydraulically rough beds are considerably higher than drag coefficients of around 0.4 for spheres at similar Reynolds number (e.g., Schlichting [29]). Other measurements made in boundary layers present results consistent with our measurements (e.g., Roberson and Chen [16], Apperley and Raudkivi [1]).

The drag coefficients we measured for seven different natural gravel particles spanned the range from lower than spheres to almost as high as cubes,

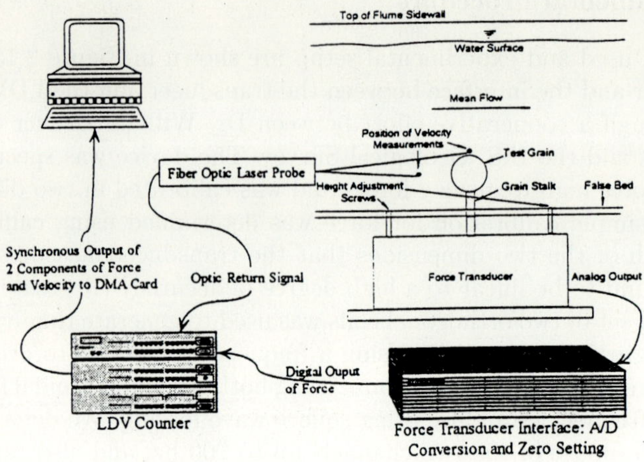

Fig. 2.13. Schematic view of flume, force transducer, and laser-Doppler velocimeter as used in experiments. Water depth is not drawn to the scale of the test grain.

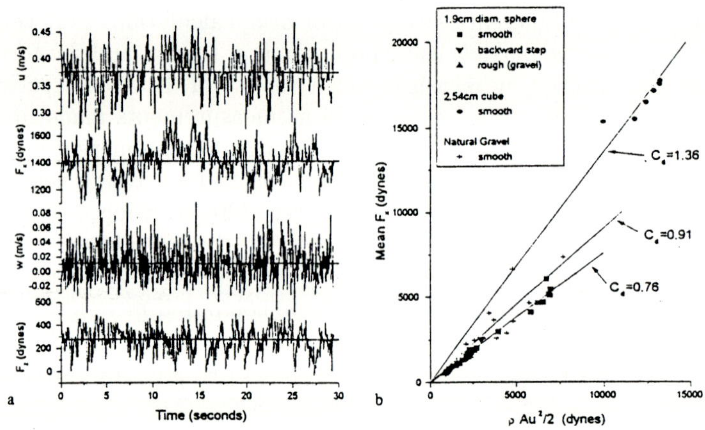

Fig. 2.14. Time series for typical experimental run. Quantities plotted from top to bottom are streamwise velocity, streamwise force, vertical velocity, and vertical force. (b) Drag coefficients for different flow conditions and particles, where A is cross-sectional area, u is average streamwise velocity, and ρ is fluid density.

depending on the shape, roughness, and orientation of the particles in the flow. The average coefficient was 0.91, significantly higher than spheres in a boundary layer, and more than double the value generally used for spheres. Drag coefficients were calculated using the squared mean velocity and the mean of the squared velocity, as well as with and without consideration of the vertical structure of the velocity over the frontal area of the grain; the conclusions about the magnitude of the coefficient were basically the same (see Schmeeckle [20]). The first model for sediment motion to incorporate these values (and the effects of turbulence) has been developed and tested by Schmeeckle [20].

Our data support the use of a standard drag closure, even for instantaneous (i.e., high-frequency) velocity, as shown by the drag coefficients plotted in Figure 2.14b. Correlation coefficients between drag and upstream horizontal velocity components were as high as 0.8, with the majority of values in the 0.6 to 0.8 range. Drag is weakly negatively correlated with w (-0.2 to -0.3, typically), as one would expect given typical negative correlations of 0.3 to 0.4 between u and w in boundary layer flows. Particle Reynolds numbers for the cases shown varied between 1000 and 10000. Lift is poorly correlated with either the velocity upstream of the grain or the velocity immediately above the grain. Nevertheless, we calculated lift coefficients, and values were comparable to those present in the literature (around 0.2), but mean lifts were typically negligible relative to instantaneous values, and computations of mean lift coefficients show wide scatter.

2.4.3 Force and turbulence structure

In Figures 2.15 and 2.16, distributions of lift and drag on a sediment particle are shown for three cases: a steady, uniform hydraulically smooth boundary layer, a developing boundary layer under a wake (20 step heights downstream of a backward step), and for a gravel bed of the same nominal particle diameter as the test grain. In the first two cases, the spherical particle was located 5mm from the bed, and for the last case the centerline of the particle was at the level of the tops of the surrounding grains; the step height was 2 particle diameters. The mean velocity was somewhat (about 25 percent) higher in the well-developed boundary layer at this level, but much higher instantaneous forces occur in the developing boundary layer and over the gravel. Maximum drag is about 25 percent greater than the mean in the simple boundary layer, over twice the mean in the developing boundary layer and even higher for the gravel. Maximum lifts in the simple boundary layer are twice the mean; they are about six times the mean in the developing boundary layer and over the gravel. The results for the backward step case show that the effect of the wake turbulence is dominant, not only near the reattachment point at 6-7 heights from the step, but even 20 step heights downstream. The effect of overlying wake turbulence has a profound effect on the boundary layer structure (Bradshaw and Wong [3], Eaton and Johnstone [5]) and on sediment transport Nelson et al. [13]. Notably, measurements of forces on a grain resting on a gravel bed show even greater variability in lift and drag than those measured 20 step heights from the backward step (Schmeeckle [20]).

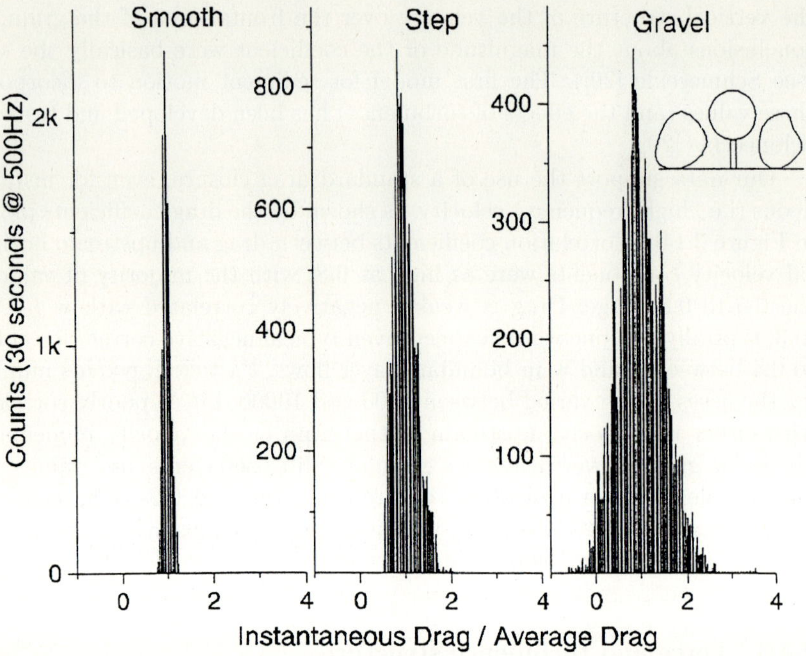

Fig. 2.15. Distributions of drag for a 1.9cm particle 5mm from the bed in a smooth turbulent boundary layer (upper panels), 5mm from the bed and 20 step heights downstream of a step 2 grain diameters high, and positioned as shown in the inset for a gravel bed.

These measurements of force show that variability in flow structure that is not captured in mean quantities such as average velocity or boundary shear stress can have a significant impact on the magnitude and frequency distribution of force on sediment grains. Thus, developing an accurate means of predicting sediment transport in complex flows requires understanding the role of turbulence more precisely than is possible from averaged quantities.

2.5 A Model for Bedload Transport in Arbitrary Flows

Although the experiments described above provide a clearer view of the coupling between flow and sediment motion, they do not in themselves provide a method for predicting transport given detailed flow information. There are two ways to use the experimental data to construct such a predictive tool. The first method is to use the velocity and sediment-transport data to construct purely empirical relations using the tools of nonlinear time series analysis (Rubin et al. [17,18]), the second is to predict transport from a physically

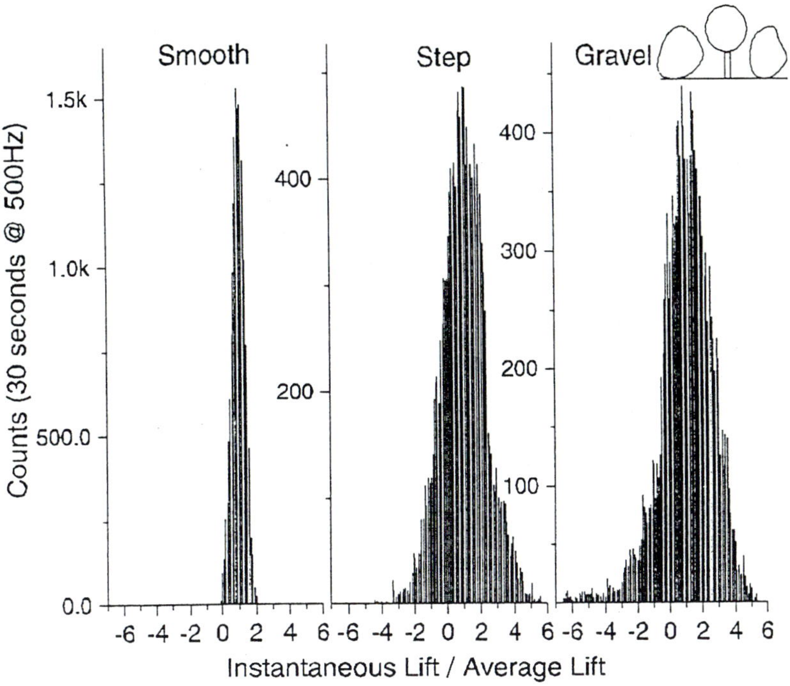

Fig. 2.16. Distributions of lift for a 1.9cm particle 5mm from the bed in a smooth turbulent boundary layer (upper panels), 5mm from the bed and 20 step heights downstream of a step 2 grain diameters high, and positioned as shown in the inset for a gravel bed.

based model using the data to set the parameters for certain components of the model (e.g., drag coefficients, lift coefficients, etc). Space will not permit a lengthy discussion of either of the two methods, but a brief overview of the physically based approach developed by Schmeeckle [20] will help to demonstrate their potential application.

Given the geometry of a sediment bed, drag coefficients for the particles making up that bed, the equations of motion for the particles, and flow forcing information, Schmeeckle [20] developed a complete approach for predicting sediment entrainment and motion as bedload. To specifically treat the mean and time-varying flow structure, the model uses a near-bed time series of horizontal and vertical velocity, and distributes that information locally by assuming that velocities can be assumed to be spatially uncorrelated for the purposes of computing instantaneous forces on particles. Particle sheltering is explicitly accounted for (see Schmeeckle [20]), as are the heights and contact points of the particles making up the bed. By numerically integrating the particle equations of motion subject to the parameterization of fluid forcing

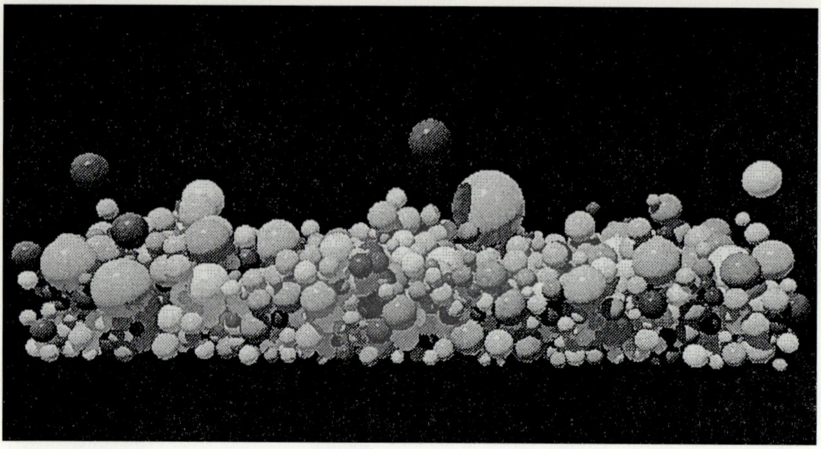

Fig. 2.17. Ray-traced image of simulated mixed grain size bed during transport. Mean grain size is 0.5cm and the length and width of the bed segment are 20 and 5cm, respectively.

obtained from the force experiments described in the previous section, the time history of every particle making up the bed can be determined. A typical "snapshot" of such a calculation is shown in Figure 2.17. For different forcing time series, the transport is readily seen to range from highly episodic to relatively consistent. Time series with the same vertical momentum transport (i.e., stress) can show very different rates of transport, consistent with our experimental measurements. In order to investigate the accuracy of this direct numerical simulation of bedload motion, we computed transport rates for a variety of stress values using a forcing time series measured over a spatially uniform gravel bed. Fluxes were computed for both single and mixed size beds, and the results were compared to the data set of Meyer Peter et al. [9] and the equations of Meyer Peter and Muller [10] and Fernandez Luque and Van Beek [6], as shown in Figure 2.18. The model performs well relative both to measured data and empirical formula when applied to steady, uniform conditions. In addition, it has general applicability to situations with unsteady and or nouniform conditions, provided that a near-bed time series is provided to infer fluid forcing on the sediment particles. Interestingly, the predictions above were made with no requirement of a so-called Bagnold or Owens boundary condition (whereby the average stress must be reduced to the critical shear stress at the boundary) because the model explicitly treats variability in the flow. Generally, our computations suggest that mean conditions are not capable of producing entrainment except at very high values of mean near-bed velocity. For steady uniform flow conditions, mean conditions appear to be insufficient for producing entrainment until the stress is almost

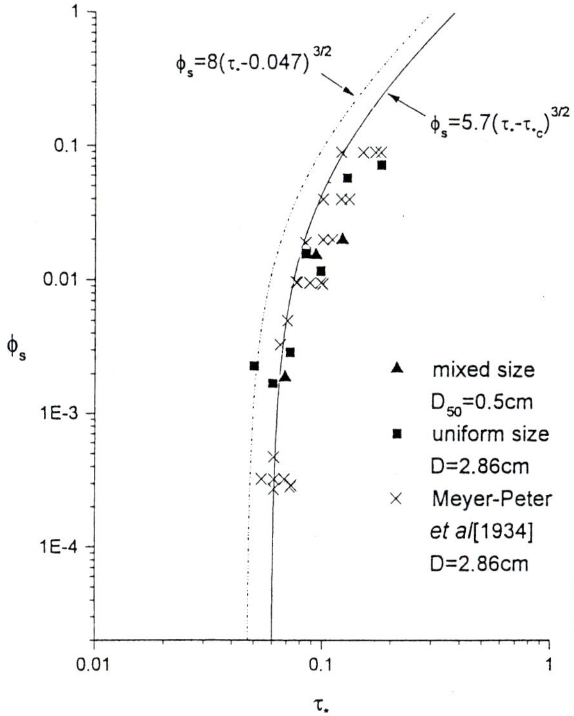

Fig. 2.18. Comparison of calculated transport with data and empirical formulas. Calculations are shown for uniform size sediment with diameter 2.86cm and for mixed-size sediment with mean diameter of 0.5cm and a lognormal distribution with a standard deviation of 0.2cm.

an order of magnitude larger than the critical shear stress. At these values, it seems that there must be some significant modification of the flow structure as a result of the presence of the sediment. However, at these high transport stages, it also appears that grain-grain interactions may be important for understanding vertical momentum flux in the fluid. Further work will be required in this area, but high concentrations are difficult to deal with using the experimental techniques employed in our previous work, so new methods must be brought to bear on this important problem. Nevertheless, the model of Schmeeckle [20] appears to work well, and combining it with either measured time series or time series computed from large-eddy or direct numerical simulations of complex flow fields is a promising avenue of further research.

2.6 Summary and Conclusions

Over the past few years, computational techniques for predicting the behavior of coupled flow-bed-sediment systems have evolved dramatically, and it is now

3 Alluvial Roughness in Streams with Dunes: A Boundary-Layer Approach

Juan J. Fedele and Marcelo H. García

V. T. Chow Hydrosystems Laboratory Department of Civil and Environmental Engineering University of Illinois at Urbana-Champaign

3.1 Abstract

The motion of sediment and development of bed forms in natural rivers still constitute intriguing phenomena that always challenge engineers and researchers involved in a variety of problems including river hydraulics and other geophysical and environmental flows. Resistance to flow in alluvial streams has been widely investigated. Since the pioneer work of Exner [12] and, later, Einstein and Barbarossa [8], the profound effects of movable sediments on the flow structure have been recognized as relevant in order to determine the nature of flow resistance (Kennedy [21]; Yalin [40]; Vanoni [36]; García and Parker [14]). This paper presents a simple methodology, based on energy balance, to compute the components of the total shear stress (grain and form-drag) acting on a uniform, two-dimensional flow over fully developed dunes. The method considers mainly an analysis of spatially-averaged (over several dune wavelengths) shear stress distributions. Then, boundary-layer theory is used in order to find an equivalent turbulent-wall-shear flow for the actual spatially-averaged flow over dunes and to derive a characteristic roughness length for the logarithmic velocity distribution above the dunes. The corresponding roughness function that also appears in the logarithmic velocity profile is also investigated and related to dimensionless sediment transport parameters. Alluvial roughness is found to be a function of a single parameter which includes the total Shields stress, a particle Reynolds number, and the ratio between flow depth and sediment diameter.

3.2 Introduction

A comprehensive description of the flow over a dune is given by Nelson et el. [25] and Bennett and Best [3]. The resulting velocity and shear stress profiles are determined by complex interactions between the propagating wake that appears after flow separation at the dune crest, the pressure gradients induced by bottom topography, and a developing internal boundary layer that grows after flow reattachment and develops until the next flow separation. Flows over dunes are naturally three dimensional as bedforms always overlap and interleave, and have finite spans, most of the cases presenting their crests aligned at a different angle than the normal respect to the flow direction. However, the idealization of two dimensionality is usually claimed in many analyses of flow over dunes since it provides a simplification that disregards what are to be identified as secondary effects (and their nonlinear interactions) on the main physics of the phenomenon.

Over the years, different procedures have been developed in order to obtain the components of the mean total shear stress, and therefore, the components of the total friction factor in a two-dimensional flow past dunes (Einstein [7]; Yalin [39]; Engelund [9,10]; Engelund and Fredsøe [11]; Vanoni and Hwang [35]; Vittori et al. [37]; Smith and McLean [32]; Klaassen [23]; Van Rijn [34]). Despite slight variations in their conceptions, most of those approaches followed the arguments of Einstein and Barbarossa [8], for shear stress partition, that is:

$$\tau_T = \tau_f + \tau_s \tag{3.1}$$

where τ_T is the total shear stress acting in the flow, τ_s the grain shear stress component, and τ_f the form-drag shear stress component. The linear superposition, eqn. (3.1), is based on the assumption that separate functions of different sets of dimensionless numbers should be found, according to the nature of the resistance introduced by each of the components (Yen [41]). It might be argued here that since the characteristic turbulence scales (length, time, etc.) associated with each source of energy dissipation in eqn. (3.1), namely wake and grain-friction (wall turbulence), are so different and also since the major source corresponds to the form-drag component, nonlinear interactions and other minor sources of dissipation (as, for instance, grain-friction or secondary flows) represent only small-departures from the leading term in eqn. (3.1). Furthermore, it has been pointed out in this regard the fact that each component is affected by different regions of the bed (Chien and Wan [5]), the difference due to non linear averaging being absorbed by the term τ_f (Yen [41]). The physical implications embedded in eqn. (3.1) are the basis for what follows below.

3.3 Laboratory Observations of Bennet and Best [3] and Theoretical Considerations

Observed local and spatially-averaged velocity and Reynolds' stress profiles show a two-layered structure of the flow over dunes (Smith and McLean [32]; Nelson et al. [25]; Bennett and Best [3]; Sukhodolow [33]). Particularly, Reynolds stress profiles present an increasing variation with depth within the layer closer to the boundary (referred hereon as internal layer), attaining a maximum at a level above which they decrease linearly with depth (referred hereon as external - or upper - layer), resembling an equilibrium boundary-layer flow. On the other hand, it is observed from laboratory experiments that spatially-averaged velocity profiles show approximately logarithmic distributions (Nelson et al. [25]; Bennett and Best [3]), with different slopes within each of those layers (which, in turn, can be used to approximate their characteristic velocity scales, see Smith and McLean [32]), presenting the change in their slopes at approximately the same level at which Reynolds stresses show their maxima. Unfortunately the lack of reliable field measurements of Reynolds stresses impedes a generalization of these considerations. However, extrapolations based on laboratory observations will be made in what follows, assuming a similar behaviour for natural, approximately two-dimensional flows over dunes.

The configuration of the spatially-averaged (over several dune wavelengths) Reynolds stress distribution resembles remarkably well those observed in turbulent shear flows with 'roughness layers' (Raupach [28]; Nezu and Nakagawa [26]; López and García [24]). For the local distributions, the locus of the shear stress maximum indicates the relevant processes in turbulence and shear production, that is wake propagation and shear-layer diffusion, and purely wall-related turbulence. This level also coincides with the location of maximum streamwise turbulent intensities. Figure 3.1 shows the evolution of local profiles of turbulent intensities, u'^2, and Reynolds' stresses, τ, along flow development, from the point of reattachment to the next flow separation, for the experiments performed by Bennett and Best [3]. Here, y is the vertical distance from the bed.

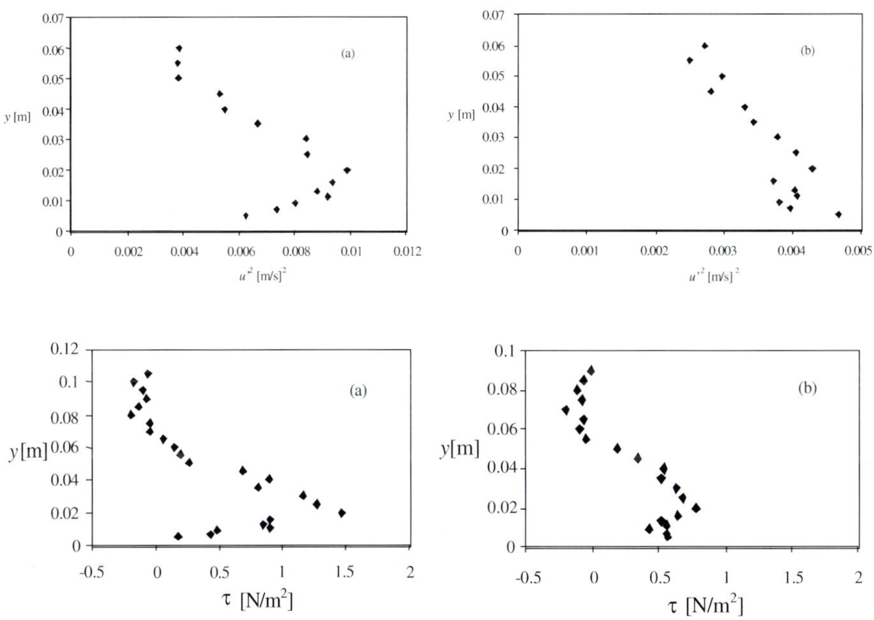

Fig. 3.1. Turbulent intensities and Reynolds stresses for (a) near reattachment, and (b) near dune crest (after Bennett and Best [3]).

Based on laboratory observations, it is now possible to propose a boundary-layer type model for the flow over a dune. A sketch of the evolution of the local Reynolds stresses, from flow reattachment to the next dune crest, is shown in Figure 3.2. This figure depicts a 'reference level' above which the Reynolds stress distribution decreases always approximately linearly with depth. We

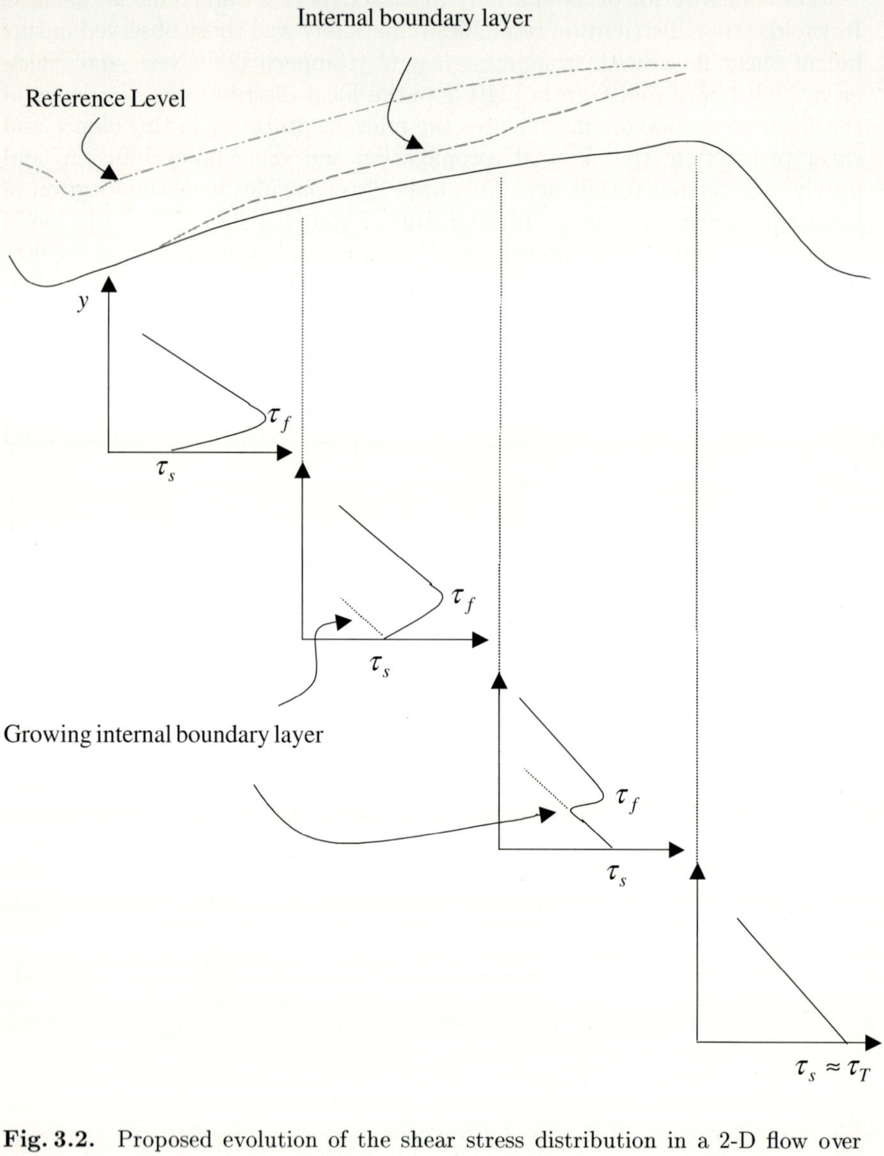

Fig. 3.2. Proposed evolution of the shear stress distribution in a 2-D flow over dunes.

shall refer to this level as 'equilibrium level' as well, since an apparent internal equilibrium, analogous to that in pure equilibrium boundary layers, is observed. On the other hand, the dynamics within the layer beneath the equilibrium level is considered here to be the result of the main mechanisms involved in mean energy expenditure, namely, wake-related losses and wall turbulence. Indeed, the migration of the shear stress maximum from a position close to the equilibrium level, in a region close to flow reattachment, to a position near the bed, in the region near flow separation at the dune crest, might indicate the main mechanisms that are competing in producing and transporting turbulent properties with varying relative importance at different locations along flow development.

It is considered herein that the maximum in the spatially-averaged Reynolds stress distribution approximates the form-drag component associated with both shear layer and averaged wake-related losses, and the extrapolation to zero depth gives the average wall shear associated with the internal boundary layer (López and García [24]). Figure 3.3 shows the Reynolds stress and velocity distributions, averaged from the reattachment point to the next downstream flow separation at the dune crest, along lines equidistant from the bottom (Nelson et al. [25]), for the experiments performed by Bennett and Best [3]). It should be noted in Figure 3.3 that the level at which Reynolds stresses attain their maximum coincides to that at which the velocity distribution changes its slope, and it is here referred to as reference or equilibrium level.

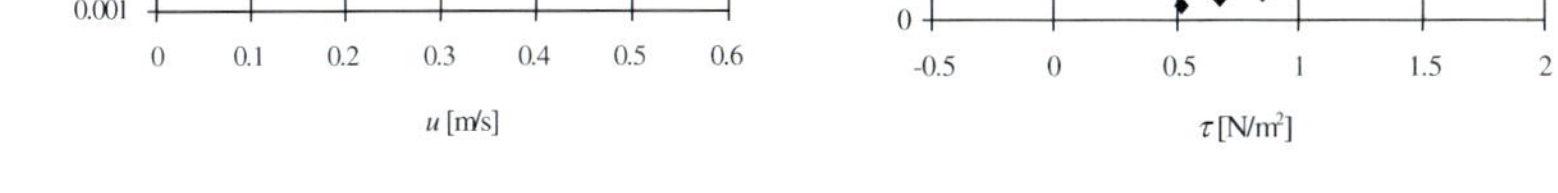

Fig. 3.3. Spatially-averaged Reynolds stress and velocity distributions (after Bennett and Best [3].

3.4 Two Layer Model for Flow over Dunes

The flow structure and shear stress distribution for flow over a dune can now be conceptualized into a two-layered flow consisting of internal and external layers.

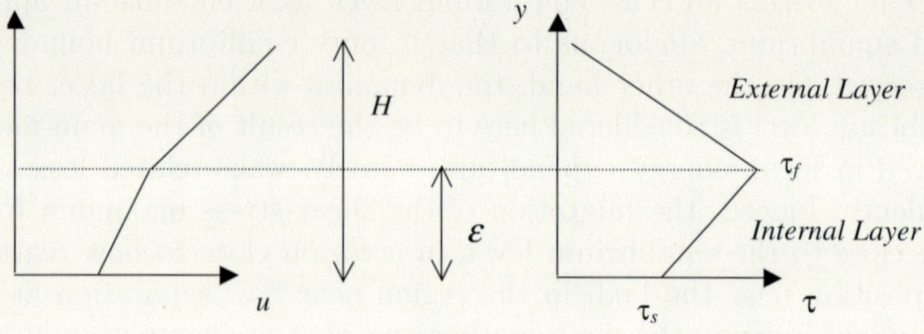

Fig. 3.4. Spatially-averaged flow over a dune.

A schematic of the spatially-averaged flow over a dune -velocity and shear stress distributions- is shown in Figure 3.4, where the mean flow depth H is considered to give the same volume of flow along a mean dune wavelength, for the actual non-uniform flow along a wavelength.

The components of the total stress acting on the flow that can be obtained from the spatially-averaged shear stress profile are depicted in Figure 3.4 as well. The maximum in this profile approximates the mean form-drag component associated with shear layer and averaged wake-related losses (form drag), at the equilibrium level ϵ, whilst the extrapolation to zero depth gives the mean wall-shear stress component (grain), associated with the internal boundary layer. The introduction of the equilibrium level ϵ represents a novel departure from the pioneer approach of Smith and McLean [32], who considered a reference level where the matching of the outer and inner velocity distributions occurred. Smith and McLean [32] also assumed that the reference level coincided with the height of the dunes. In the approach proposed herein, the equilibrium level ϵ does not necessarily have to coincide with the dune height, as will be shown below.

The form drag coefficient C_D is then investigated using both the spatially-averaged value of the form-drag component as given in Figure 3.3, and the one obtained by integrating the distribution of local values from measured local profiles (visually observed at local equilibrium levels), according to the idea depicted in Figure 3.2. In any case, a value of $C_D = 0.21$ was obtained (see Nelson et al. [25]), whereas a value of $C_D = 0.29$ was obtained using the expression proposed by Shen et al. [30] for the same spatially-averaged flow. Moreover, the observed decay with distance of the assumed local form-drag component was found to follow a power law with an exponent close to -0.8 (see Nelson et al. [25]). Now, following Smith and McLean [32], scaling within the external layer (above the equilibrium level ϵ) is accomplished by using a total shear stress resulting from the summation of components obtained as explained before. Figure 3.5 shows the distributions of streamwise turbulent fluctuations from flow reattachment to next flow separation, where the local total shear velocity u_{*T} (computed using total shear stress), and the local depth h (strictly, distance from the bed at which measured Reynolds stresses vanish) are used as velocit y and length scales, respectively. The fit in Figure 3.5 is given b y:

$$\frac{u'}{u_{*T}} = 2.04 \exp\left[-0.7\frac{(y-\epsilon)}{h}\right] \qquad (3.2)$$

here, the value $D_u = 2.04$ is in close agreement with that given by Nezu and Nakaga w a [26] for boundary layers o ver fully rough walls. However, the

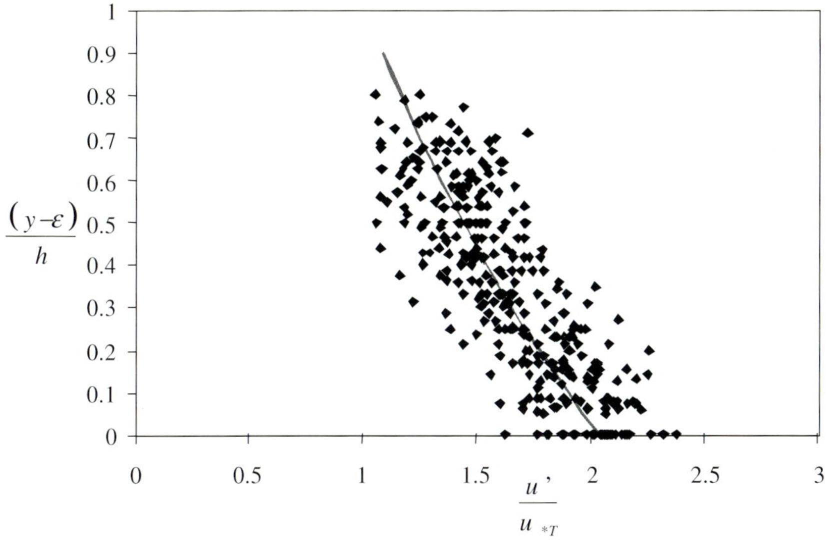

Fig. 3.5. Streamwise turbulent fuctuations for a flow past dune.

coefficient $C_k = -0.7$ reveals a slo w er decay with depth. Scaling within the internal lay er (close to the bed) is not inv estigated here in detail. How ev er, it is w orth mentioning that, even though when spatially-av eraged velocities might show an almost logarithmic distribution within the in ternal la y er (at least for the experimental data considered heretofore, see Figure 3.3), no proper scaling of a pure boundary-lay er type was found when analyzing streamwise turbulent fluctuations within this lay er. We argue that this might be related to the fact that, within the internal lay er, tw o major dynamic mechanisms are competing in producing and transporting turbulent properties with different relativ e importance at different locations along flow development (shear lay er and wak e diffusion, and wall turbulence, see Figure 3.2).

Finally, it is assumed that the velocit y at the equilibrium level ϵ is given approximately by the mean velocit y of the flow U. If the friction coefficient remains approximately constant within the external lay er (and tends to the total friction coefficient), and if the equilibrium level indeed separates tw o

Moreover, it is assumed that a fully-rough velocity distribution is valid within the internal layer, then:

$$U = \frac{u_{*s}}{\kappa} \ln\left(30\frac{\epsilon}{k_s}\right) \tag{3.10}$$

Since the condition of internal equilibrium at ϵ requires a force balance at this level, then the following relation needs to be satisfied:

$$\int_0^{\epsilon} \tau(y)dy = \int_{\epsilon}^{H} \tau(y)dy \tag{3.11}$$

Eqns. (3.9), (3.10) and (3.11) constitute a closed system for the unknowns τ_s, ϵ and k_s, that can be solved when the values of specific flow discharge, mean flow depth, dune geometry, and water surface slope are known. The value of the form-drag component is assumed to be known and computed using mean flow velocity, mean dune geometry and a constant drag coefficient $C_D = 0.21$ (Nelson et al. [25]). Finally, the total shear stress can be obtained using eqn. (3.1) once the above system is solved.

3.6 Computation of Composite Roughness Length

The above methodology was applied using laboratory flume measurements for the cases of fully developed dunes, with only one relevant bedform (no superimposed, secondary bedforms), and negligible suspended sediment transported (Guy et al. [15]; Znamenskaya [42]; Hung and Shen [18]; Wijbenga and Klaassen [50]; Smith and Ettema [31]), together with some natural river data (HRS [17]; Julien [20]) under similar flow and bed conditions. Sediment sizes D and relative roughness parameter, H/D, ranged from 0.19 to 1.12 mm, and from 150 to 9×10^4, respectively, for the data used here. Computed discharges obtained after integration of the logarithmic velocity distributions according to the parameters found in each layer showed good agreement with the reported measured ones (Fedele [13]). A composite roughness length k_c can then be computed considering the continuity of the velocity profile at the equilibrium level ϵ, for a fully rough logarithmic velocity distribution within the upper layer, that is:

$$U = \frac{u_{*T}}{\kappa} \ln\left(30\frac{\epsilon}{k_c}\right) \tag{3.12}$$

Once all the characteristic parameters defining the spatially-averaged shear stress distribution are known, and using simple geometric considerations, an equivalent depth H' can then be obtained after comparing areas given by the actual shear stress profile and a hypothetical triangular distribution (as in a turbulent-wall-shear flow) having a total shear stress equal to that present in the actual flow. Furthermore, an alternative means to compute the composite characteristic roughness length k_c is given by the use of the Moody diagram or Brownlie's [4] expression, once the total shear stress and the equivalent depth H' are known, for the same mean flow velocity U. Figure 3.6 shows the transformation of the actual force diagram into a hypothetical, equivalent pure boundary-layer one and their characteristic parameters.

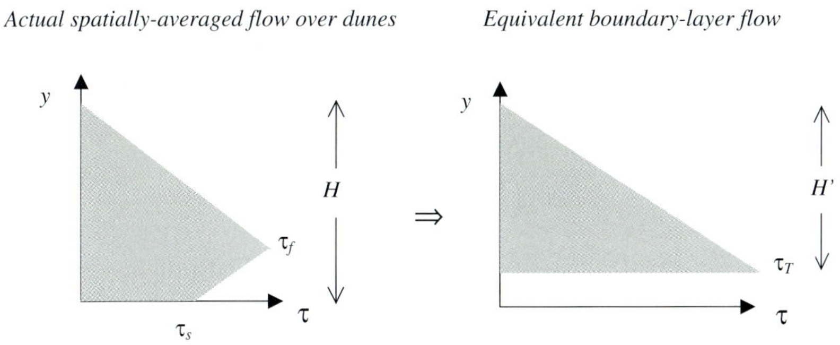

Fig. 3.6. Equivalence of total resistive force.

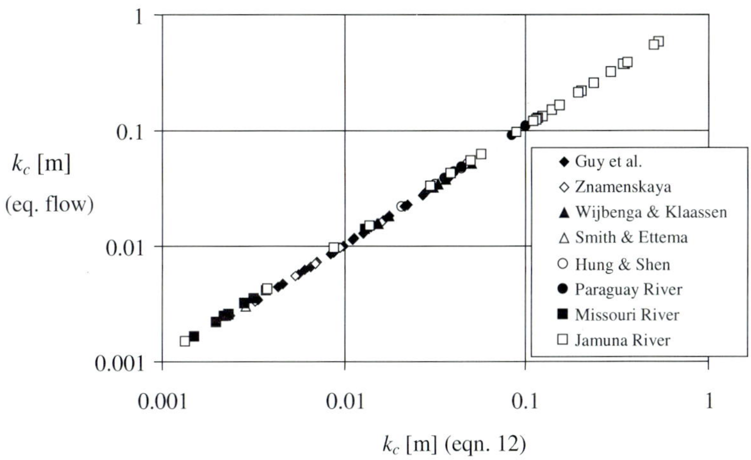

Fig. 3.7. Composite-characteristic roughness length.

It is shown in Figure 3.7 that the composite roughness length obtained by eqn. (3.12) corresponds to the one of an equivalent turbulent-wall-shear flow (named equivalent boundary-layer flow in this figure), with the same total resistance and mean velocity, but with depth H'.

The difference $\Delta H = H - H'$ is considered here to be analogous to the concept of displacement height, an additional characteristic length appearing in the logarithmic velocity distribution for fully rough walls, and proportional to the total force distribution, as presented by Jackson [19]. If this is the case, then ΔH gives the virtual origin of the equivalent boundary-layer flow, with respect to the mean flow depth of the actual alluvial flow over dunes.

3.7 Roughness Function for Alluvial Streams

The roughness function (Hama [16]; Schlichting [29]) that appears in the velocity profile proposed by Clauser [6], eqn. (3.13), can be investigated in order to define the nature of 'alluvial roughness' for flows over dunes:

$$\frac{u}{u_{*T}} = \frac{1}{\kappa} \ln\left(\frac{u_{*T}y}{\nu}\right) + A - \frac{\Delta u}{u_{*T}}\left[\frac{u_{*T}k_c}{\nu}\right] \tag{3.13}$$

here, A is a universal constant, $\Delta u/u_{*T}$ is the so-called roughness function, and ν is the kinematic viscosity of the fluid. Hama [16] has expressed a universal form for the roughness function for fully rough walls as:

$$\frac{\Delta u}{u_{*T}} = \frac{1}{\kappa} \ln\left(\frac{k_c u_{*T}}{\nu}\right) + C_k = \left(\sqrt{\frac{2}{C_f}}\right)_{smooth} - \left(\sqrt{\frac{2}{C_f}}\right)_{rough} \tag{3.14}$$

being the constant C_k a function essentially of the nature and geometry of the (fixed) roughness elements, for the $'k'$ roughness type that has been defined by Perry et al. [27], for sand covered (flat) walls, and sparse roughness elements. The roughness function (Hama [16]; Schlichting [29]) for an alluvial stream is obtained here using a methodology depicted in Perry et al. [27], which is based on the ratio of the Reynolds number of the flow (with the equivalent depth H') and the one corresponding to a smooth wall, for the same total friction coefficient f, and the computed total composite roughness length k_c (with the equivalent depth H'). Figure 3.8 shows the roughness function $\Delta u/u_{*T}$ as a function of the parameter $k_c u_{*T}/\nu$ for the laboratory experiments and field streams considered herein.

A relationship in the form of eqn. (3.14), namely the fully-rough-wall regime, can be obtained from Figure 3.8 for values of $k_c u_{*T}/\nu > 200 - 250$, and is given by:

$$\frac{\Delta u}{u_{*T}} = 2.44 \ln\left(\frac{k_c u_{*T}}{\nu}\right) - 3.24 \tag{3.15}$$

To the authors' knowledge, this is the first roughness function ever proposed to characterize alluvial roughness in the presence of dunes. A more general form of eqn. (3.15) is proposed below.

3.8 Composite Roughness and Sediment Transport

The practical benefits embedded in eqn. (3.15) rely in its capability of summarizing the effect of total resistance by a single length scale, k_c , which results

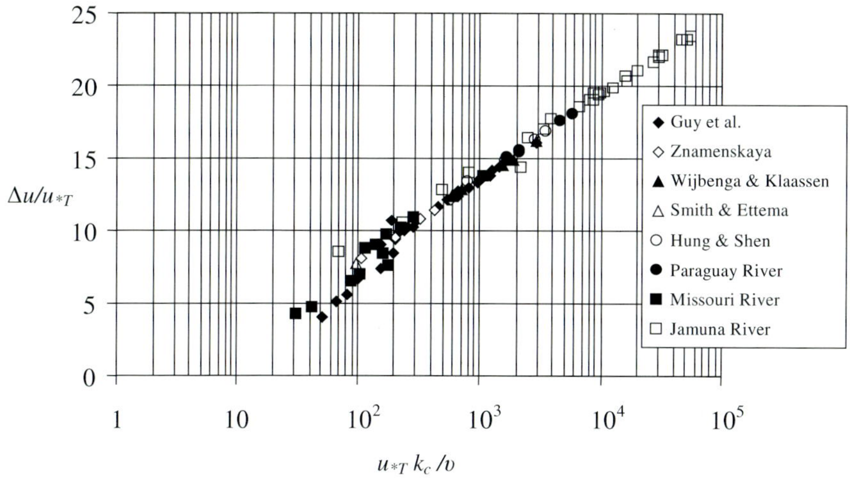

Fig. 3.8. Roughness function for alluvial streams with dunes.

directly proportional to the magnitude of the total resistive force (Jackson [19]). However, for the case of alluvial streams, this magnitude is no longer a characteristic geometric measure of the roughness elements constituting the boundary alone; rather, it represents a dynamic length scale that includes the combined effects of having a mobile bed and, therefore, a more complex, dynamic quantity to be determined when analyzing total resistance.

Thus, as the nature of the roughness in alluvial streams is determined by the interaction fluid-sediment, a general relationship betw een the roughness function and the sediment transport dimensionless numbers (Yalin [40]; Garc ía and Park er [14]) shall be given in the following general form:

$$\frac{\Delta u}{u_{*T}} = \Phi\left[\frac{u_{*T}}{v_s}; R_{ep}; \frac{H}{D}\right] \tag{3.16}$$

where $R_{ep} = \sqrt{gRD}D/\nu$ is the particle Reynolds n umber, v_s is the terminal fall velocit y of the characteristic sediment size, D, g is the gravitational acceleration, and R is the submerged specific gravit y of the sediment. Eqn. (3.16) can also be written as:

$$\frac{\Delta u}{u_{*T}} = \Phi\left[\tau_{*T}; R_{ep}; \frac{H}{D}\right] \tag{3.17}$$

since for any case, the sediment fall v elocit y can be expressed as $v_s = \sqrt{gRD}\Phi[R_{ep}]$.

Figure 3.9 shows a first attempt to finding a relationship of the form of eqn. (3.17), based on dimensionless sedimentologic parameters and for all

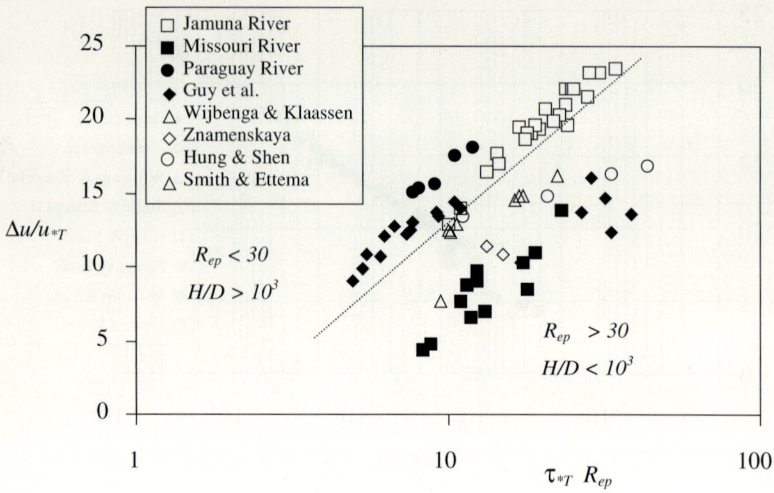

Fig. 3.9. Alluvial roughness function and sediment transport dimensionless numbers.

cases when $k_c u_{*T}/\nu > 250$ (fully rough regime, according to Figure 3.8). Iterative trials have only produced plots of the type of Figure 3.9, and no single combination of the dimensionless numbers has been found to collapse all the data within a single expression as in eqn. (3.14). After inspection of Figure 3.9, two main regions have been identified, in principle given mainly by a combination of the parameters R_{ep} and H/D, namely:

$$\frac{\Delta u}{u_{*T}} = C_1 \ln(\tau_{*T} R_{ep}) + C_2 \qquad \text{for } R_{ep} < 30 \text{ , and } H/D > 10^3 \tag{3.18}$$

where C_1 and C_2 are constants. The general case given by eqn. (3.17) should hold when $R_{ep} > 30$ and when $H/D < 10^3$. It should be noticed that the Missouri River data do not fall within the region given by eqn. (3.18) even though for this river $R_{ep} \approx 13$. There is no rational explanation for the unexpected behaviour of the Missouri River data, other than the fact that temperature effects, known for their prominent role in this river, could be at play.

The existence of different 'regions' within the fully-rough regime for alluvial streams with dunes is considered herein a result of the complex, nonlinear behaviour of the sediment motion under different conditions of characteristic sizes of sediment and flow scales, expressed by the dimensionless numbers given in eqn. (3.17).

By comparison of eqn. (3.15) and eqn.(3.16), the following general expression is proposed:

$$\frac{u_{*T} k_c}{\nu} \propto \tau_{*T}^{a} \; R_{ep}^{b} \left(\frac{H}{D}\right)^{c} \tag{3.19}$$

An iterativ emethod was used in order to obtain the exponents a, b and c, by assuming expressions of the form of eqn. (3.14), and upon substitution of eqn. (3.19) into the argument of eqn. (3.14). The values of the pow ers were considered satisfactory when the obtained fit gave a slope represented by the inv erse of the von Karman constant (≈ 2.5).

F urthermore, each of these expressions should match one into another as the c haracter of transport of sediment and total roughness change. Follo wing the above arguments, it w as found that:

$$\frac{\Delta u}{u_{*T}} = \frac{1}{\kappa} \ln \left(\tau_{*T}^2 R_{ep}^2 \left(\frac{H}{D} \right)^{1/3} \right) + C_{Tr} \quad \text{for } R_{ep} < 30 \text{ , and } H/D > 10^3 \tag{3.20}$$

and

$$\frac{\Delta u}{u_{*T}} = \frac{1}{\kappa} \ln \left(\tau_{*T}^{1/2} R_{ep}^{1/2} \left(\frac{H}{D} \right)^2 \right) + C'_{Tr} \quad \text{for } R_{ep} > 30 \text{ , and } H/D < 10^3 \tag{3.21}$$

as tw o extreme cases, say, and the constants C_{Tr} and C'_{Tr} account for all the transformations betw een the arguments of eqn. (3.14), and eqns. (3.20) and (3.21), for each region.

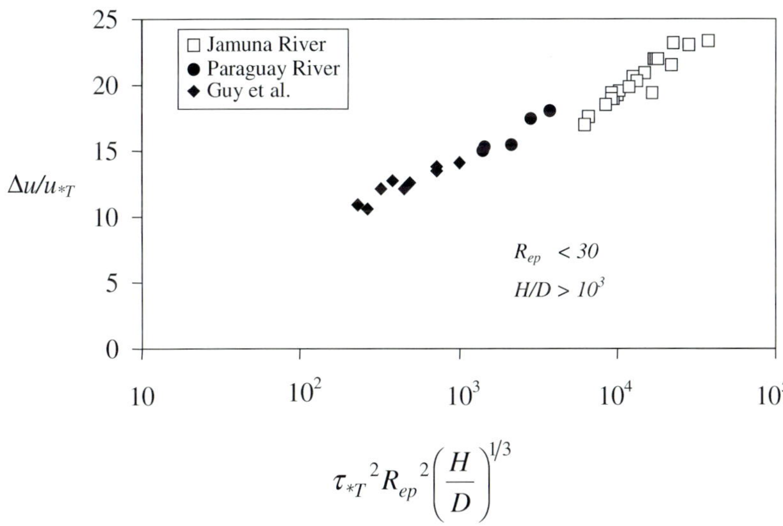

Fig. 3.10. Roughness function, eqn. (3.20). (Missour River data are not included).

Figure 3.10 shows a relationship of the form of eqn. (3.20) for which the best fit resulted in:

$$\frac{\Delta u}{u_{*T}} = 2.39 \ln \left(\tau_{*T}^2 R_{ep}^2 \left(\frac{H}{D} \right)^{1/3} \right) - 2.34 \tag{3.22}$$

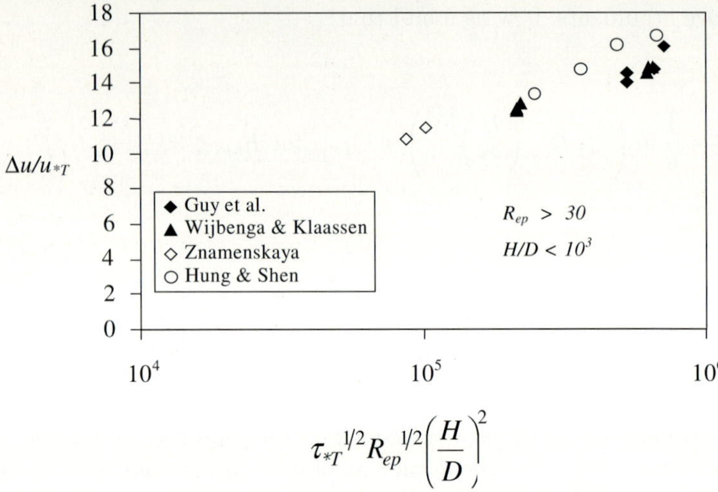

Fig. 3.11. Roughness function (eqn 3.21).

Figure 3.11 shows a relationship of the form of eqn. (3.21), for which the best fit has resulted in the following expression:

$$\frac{\Delta u}{u_{*T}} = 2.2 \ln \left(\tau_{*T}^{1/2} R_{ep}^{1/2} \left(\frac{H}{D} \right)^2 \right) - 13.5 \tag{3.23}$$

Finally, by comparison of the arguments of eqn. (3.14), and eqns. (3.20) and (3.21), relationships for the total roughness length k_c, based on dimensionless numbers, are found. Figure 3.12 presents one of such relationships, for the case given b y eqn. (3.20), whereas Figure 3.13 gives the case represented b y eqn. (3.21). It is observed from Figure 3.13 that for high values of R_{ep}, the total roughness length can apparently be expressed as a function mainly of the relative roughness H/D.

Now, it is suggested that a general expression for the roughness function can be written in the following form:

$$\frac{\Delta u}{u_{*T}} = \frac{1}{\kappa} \ln \left(\tau_{*T}^{1/2} R_{ep}^{1/2} \left(\frac{H}{D} \right)^2 \right) + \frac{\Delta u^*}{u_{*T}} [\Phi_{Tr}] \tag{3.24}$$

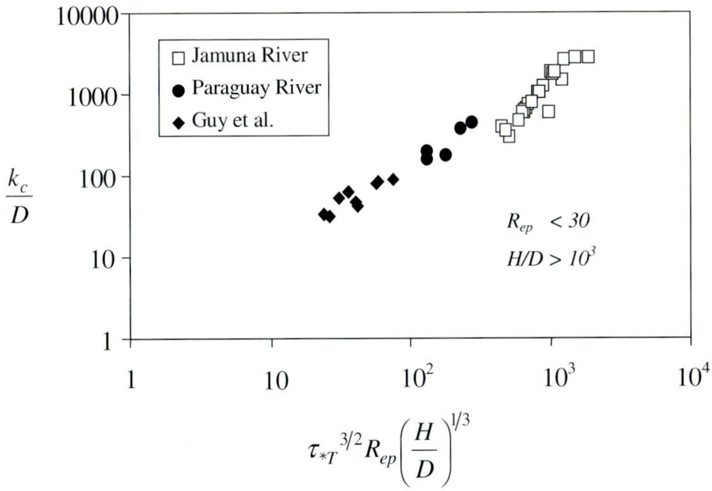

Fig. 3.12. Total roughness length for flow with dunes, from eqn. (3.20).

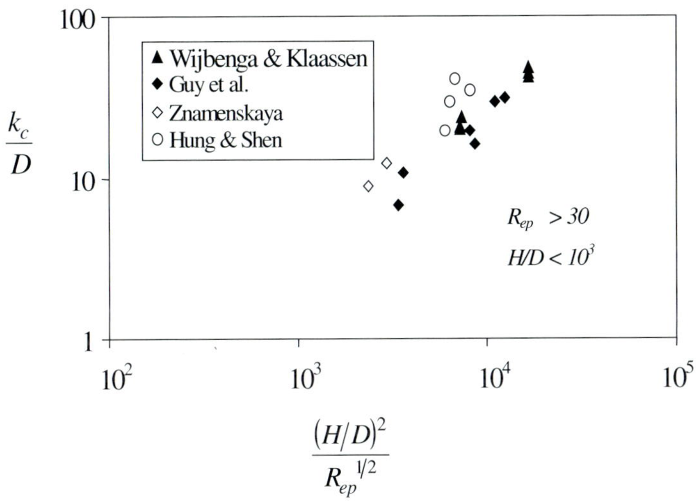

Fig. 3.13. Total roughness lenght for flow with dunes, from eqn. (3.21).

In eqn. (3.24), the existence of different regions determining different behaviour of the sediment transport phenomenon, and therefore its relevance on determining the nature of the total roughness, is expressed by an additional term for the roughness function, $\Delta u^*/u_{*T}$, and given by:

$$\frac{\Delta u^*}{u_{*T}} = f[\Phi_{Tr}] \tag{3.25a}$$

where

$$\Phi_{Tr} = \frac{\tau_{*T}^{3/2} R_{ep}^{3/2}}{\left(\frac{H}{D}\right)^{5/3}} \tag{3.25b}$$

with the following limiting cases:

$$\frac{\Delta u^*}{u_{*T}} \propto \frac{1}{\kappa} \ln(\Phi_{Tr}) \qquad \text{for } \Phi_{Tr} \to 0 \tag{3.26a}$$

and

$$\frac{\Delta u^*}{u_{*T}} \to C'_{Tr}(const.) \qquad \text{for } \Phi_{Tr} \to 0.01 \tag{3.26b}$$

here, a roughness-transport parameter for fully-rough flo ws with fully developed dunes is given by Φ_{Tr}, which is considered to express mainly scale effects on the relationship between sediment transport and the distribution of the total resistance forces, represented by the abo ve metioned 'regions' (see Figures 3.10 and 3.11).

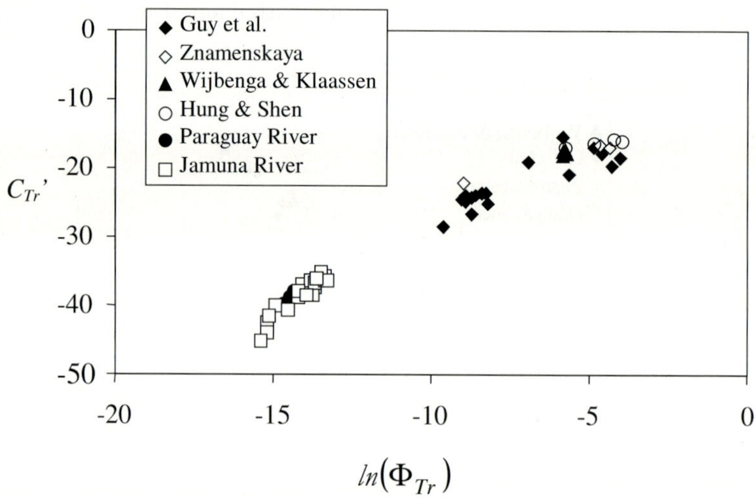

Fig. 3.14. Roughness-transport parameter for flows with fully developed dunes.

The limiting cases, eqns. (3.26a, 3.26b), are shown in Figure 3.14, where the v ariation of the constant C'_{Tr} with $\ln(\Phi_{Tr})$ is giv en and:

$$C'_{Tr} = \frac{\Delta u}{u_{*T}} - \frac{1}{\kappa} \ln\left(\tau_{*T}^{1/2} R_{ep}^{1/2} \left(\frac{H}{D}\right)^2\right) \tag{3.27}$$

It is necessary to point out here that, according to what is presented, expressions for total friction coefficients or v elocit y distributions of the boundary-lay er type ha veto include the parameter H' that has been obtained from the equivalence between the actual spatially-averaged flow over dunes and a turbulent-wall-shear flow.

3.9 Total Friction Coefficient

Now the connection between the roughness function presented above, and the commonly used total friction coefficient is sought. The total friction coefficient for fully-rough walls is expressed as (Keulegan [22]):

$$\frac{1}{\sqrt{C_f}} = \frac{1}{\kappa} \ln\left(\frac{H'}{k_c}\right) + const \tag{3.28}$$

Using the dimensionless relations for k_c shown in Figures 3.12 and 3.13, and considering that $H'/H \cong$ constant (Jackson [19]), for this type of roughness (Fedele [13]), expressions for the total friction coefficient based on dimensionless numbers, for the different regions, according to the limiting values of Φ_{Tr}, were obtained as:

$$\frac{1}{\sqrt{C_f}} = \frac{1}{\kappa} \ln\left(\frac{(H/D)^{2/3}}{\tau_{*T}^{3/2} R_{ep}}\right) + const. \qquad \text{for } \Phi_{Tr} \to 0 \tag{3.29a}$$

and

$$\frac{1}{\sqrt{C_f}} = \frac{1}{\kappa} \ln\left(\frac{R_{ep}^{1/2}}{H/D}\right) + const. \qquad \text{for } \Phi_{Tr} \to 0.01 \tag{3.29b}$$

where, analogously to what was presented above for the roughness function, the limiting cases here are matched by using the parameter Φ_{Tr}, as well. That is:

$$c_{Tr} = \frac{1}{\sqrt{C_f}} - \frac{1}{\kappa} \ln\left(\frac{R_{ep}^{1/2}}{H/D}\right) \propto \ln\left(\frac{1}{\Phi_{Tr}}\right) \qquad \text{for } \Phi_{Tr} \to 0 \tag{3.30a}$$

and

$$c_{Tr} = \frac{1}{\sqrt{C_f}} - \frac{1}{\kappa} \ln\left(\frac{R_{ep}^{1/2}}{H/D}\right) \to const. \qquad \text{for } \Phi_{Tr} \to 0.01 \tag{3.30b}$$

Figure 3.15 shows the corresponding variation of c_{Tr} with $\ln(1/\Phi_{Tr})$, as expected from direct consequence of the relationship given by eqn. (3.24).

Figures 3.16 and 3.17 show the proposed relationships for the total friction coefficient for the limiting cases $\Phi_{Tr} \to 0$, and $\Phi_{Tr} \to 0.01$, respectively.

From the above analysis, it follows that the total friction coefficient in alluvial streams with fully-developed dunes is a function of the total Shield stress, sediment size, and relative flow depth. This dependence is highly non linear and is also non unique for the fully-rough-wall regime. Yalin [40] (p. 270) did point out the effect of particle size and relative flow depth on total friction coefficients. Here, boundary-layer theory (Hama [16]) is used to show why this is the case.

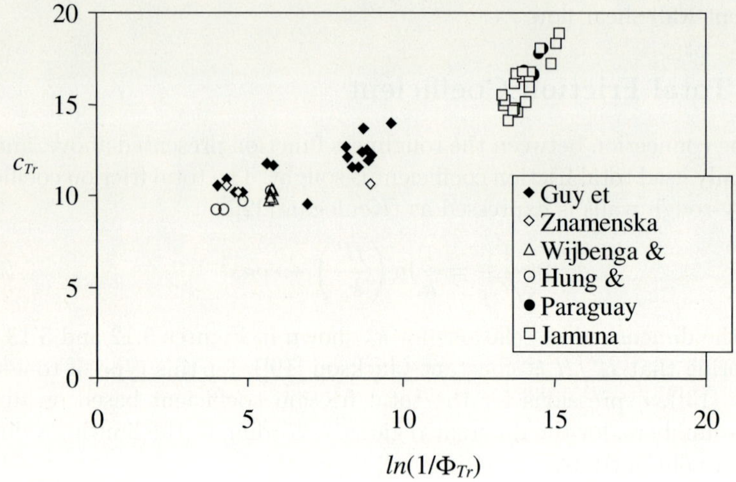

Fig. 3.15. Roughness transport parameter and total friction coefficient.

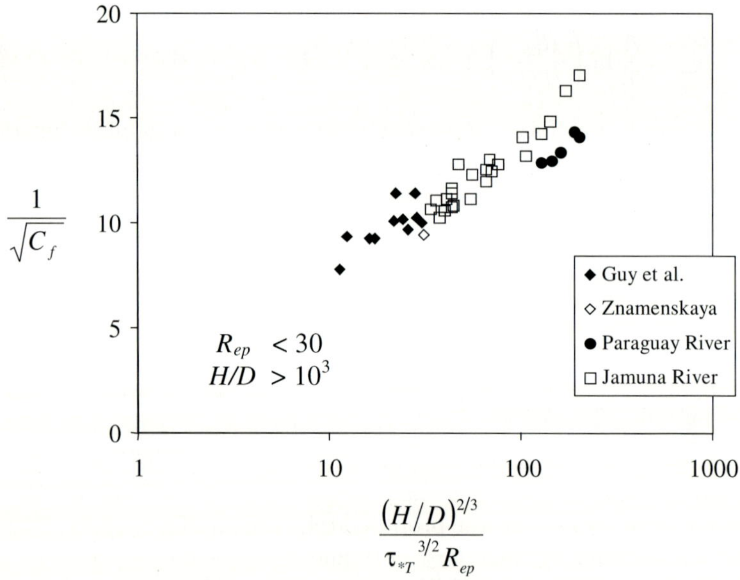

Fig. 3.16. Total friction coefficient for $\Phi_{tr} \to 0$.

3.10 Conclusions

A methodology to determine shear stress partition for flows with equilibrium, fully-developed dunes, based on simple energy considerations, is presented.

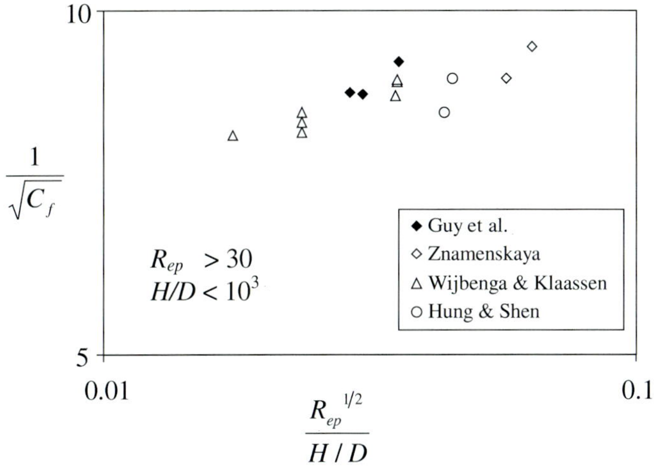

Fig. 3.17. Total friction coefficient for $\Phi_{T_r} \to 0.01$.

It is argued that an 'equivalent depth' is needed for making use of the traditional friction coefficients that have been developed for the case of equilibrium, boundary-layer flows. This is based on a resistance force equivalence between the actual spatially-averaged flow and a hypothetical pure boundary layer one, having both the same mean flow velocity. On the other hand, it is considered that the complex interactions of wake propagation, pressure gradients and internal boundary layer can be regarded to be confined within an internal layer close to the bottom where the major energy expenditure occurs. Above this internal layer, an internal equilibrium is reached in the flow manifested in a logarithmic velocity distribution and a linearly decreasing shear stress distribution. It is shown that the level at which the spatially-averaged Reynolds stress profiles attain a maximum (characteristic value of form-drag losses) can be regarded as an equilibrium level with respect to which internal forces are balanced, dividing two different dynamic regions.

The roughness function for alluvial streams with dunes is then investigated for a wide range of stream scales, suggesting a fully rough regime for values of $k_c u_{*T}/\nu > 200 - 250$. Finally, the results of dimensional analysis for sediment transport are included and a combined approach between both theories, sediment transport and the classical boundary layer one is suggested by expressing the roughness function in terms of dimensionless numbers for sediment transport. Dimensionless functions for the composite roughness length k_c are then derived from the roughness function for flow over dunes. It is found that scale effects on total resistance and sediment transport can be represented by a single parameter involving the total Shield stress (τ_{*T}), the particle Reynolds number (R_{ep}), and the relative flow depth (H/D).

More field and laboratory data are needed for flows over dunes, in particular to achieve a better understanding of the flow structure within the inner layer. Until these observations become available, it will not be possible to improve our understanding of the physics of the flow. Fortunately, there is now technology such as acoustic Doppler profilers that can be used to assess the flow structure as well as the sediment concentration field above dunes (Admiral and García [1])

3.11 Acknowledgements

The authors wish to thank Jim Best and Sean Bennett for providing the data from their laboratory experiments on flow over dunes. This research is a by-product of work conducted with support from the Marine Geology and Geophysics Program of the Office of Naval Research (ONR) and the Water and Watersheds Program of the US Environmental Protection Agency (EPA). This financial support is gratefully acknowledged.

References

1. Admiral, D.M., García, M.H. (2000) Laboratory measurement of suspended sediment concentration using an Acoustic Concentration Profiler (ACP). Experiments in Fluids, **28**, 116-127.
2. Amsler, M.L., Schreider, M.I. (1992) Caractersticas hidráulicas de la superposición de formas de fondo en el río Paraná (Argentina). Proc. XV Latin American Congress, IAHR, Vol.3, Cartagena, Colombia, 39-48, (in Spanish).
3. Bennett, S., Best, J. (1995) Mean flow and turbulence structure over fixed, two-dimensional dunes: Implications for sediment transport and bedform stability. Int. As. of Sedimentologists, Sedimentology, **42**, 491-513.
4. Brownlie, W. (1981) Re-examination of Nikuradse roughness data. Proceedings of ASCE, Vol. 107. Technical Notes.
5. Chien, N., Wan, Z. (1999) Mechanics of sediment transport. ASCE Press, USA.
6. Clauser, F.H. (1954) Turbulent boundary layers in adverse pressure gradients. J. Aeronautical Sciences, **21**, 91.
7. Einstein, H.A. (1950) The bed-load function for sediment transportation in open channel flows. Technical Bulletin No 1026, U.S. Dept. of Agri., Washington D.C.
8. Einstein, H.A., Barbarossa, N. (1952) River channel roughness. Trans. ASCE **117**, 1121-1146.
9. Engelund, F. (1967) A sediment transport theory based on similarity. Hydraulic Lab., Tech. Univ. of Denmark, Basic Research Progress Report 13.
10. Engelund, F. (1977) Hydraulic resistance for flow over dunes. Inst. Hydrodyn. and Hydraulic Engrg. Tech. Univ. Denmark, Proc. Rep. 44, 19-20.
11. Engelund, F., Fredsøe, J. (1982) Sediment ripples and dunes. Annual Review of Fluid Mechanics, **14**, 13-37.
12. Exner, F.M. (1925) Uber die Wechselwirkung zwischen Wasser und Geschiebe in Flussen. Sitzber Akad. Wiss, 165-180. (Translation).

13. Fedele, J.J. (1999) Bed roughness in alluvial streams, Ms Thesis, University of Illinois, 111.
14. García, M.H., Parker, G. (1991) Entrainment of bed sediment into suspension. J. Hyrd. Engin., ASCE **117** 4, 414-435.
15. Guy, H.P., Simons, D.B., Richardson, E.V. (1966) Summary of alluvial channel data from flume experiments 1956-61. Prof. Paper 462-I, US Geological Survey.
16. Hama, F. (1954) Boundary layer characteristics for smooth and rough surfaces. Trans. Society of Naval Architects and Marine Engineers, USA **62**, 333-351.
17. HRS (Hydraulic Research Station). (1972) River Paraguay Study. Report No Ex 606, Wallingford, Berkshire, U.K.
18. Hung, C.S., Shen, H.W. (1979) Statistical analysis of sediment motion on dunes. J. Hydr. Eng., ASCE, **105**(3), 590-610.
19. Jackson, P. (1981) On the displacement height in the logarithmic velocity profile. J. Fluid Mechanics, **111**, 15-25.
20. Julien, P.Y. (1992) Study o bed form geometry in large rivers. Delft Hydraulic Lab.
21. Kennedy, J.F. (1963) The mechanics of dunes and antidunes in erodible-bed channels. Journal of Fluid Mechanics, **16**, 521-544.
22. Keulegan, G.H. (1938) Laws of turbulent flow in open channels. J. National Bureau of Standards, Research paper 1151, **21**, 707-741.
23. Klaassen, G.J. (1979) Sediment transport and hydraulic roughness in relation to bed forms. Delft Hydraulic Laboratory, Publ. No 213.
24. López, F., García, M.H. (1999) Wall similarity in turbulent open-channel flow. J. Engineering Mech., **125**(7).
25. Nelson, J., McLean, S., Wolfe, S. (1993) Mean flow and turbulence fields over two-dimensional bed forms. Water Resources Research, **29**(12), 3935-3953.
26. Nezu, I., Nakagawa, H. (1993) Turbulence in open-channel flows. IAHR Monograph Series, A.A. Balkema, Rotterdam.
27. Perry, A., Schofield, W., Joubert, P. (1969) Rough wall turbulent boundary layers. J. Fluid Mechanics, **37**, parte 2, 383-413.
28. Raupach, M.R. (1981) Conditional statistics of Reynolds stress in rough-wall and smooth-wall turbulent boundary layers. J. Fluid Mech., **108**, 363-382.
29. Schlichting, H. (1979) Boundary-layer Theory. McGraw Hill Edit., 7th edition.
30. Shen, H.W., Fehlman, H., Mendoza, C. (1990) Bed form resistances in open channel flows. J. Hydr. Engin., **116**(6), 799-815.
31. Smith, B.T., Ettema, R. (1994) Floating cover influence on sediment transport in dune-bed channels. Hydraulic Engineering '94, ASCE, 1156-1160.
32. Smith, J.D., and McLean, S. (1977) Spatially averaged flow over a wavy surface. J. Geophysical Research, **82**(12), 1735-1746.
33. Sukhodolov, A., Thiele, M., Bungartz, H. (1998). Turbulence structure in a river reach with sand bed. Water Resources Research, **5**, 1317-1334.
34. van Rijn, L. (1982) The prediction of bed forms, alluvial roughness and sediment transport. Research Report, Delft Hydraulics Laboratory, S484, Part III.
35. Vanoni, V.A., Hwang, L.S. (1967) Relation between bed forms and friction in streams. J. Hydr. Division, ASCE, **100**(HY3), 363-377.
36. Vanoni, V.A. (1979) Effect of dunes on sand bed streams, Tech. Memorandum 79-3, California Institute of Technology, Pasadena, California. Presented at the 3rd Brazilian Symposium on Hydrology, Brasilia, Aug. 1979.
37. Vittal, V.A., Ranga Raju, K.G., Garde, R.J. (1977) Resistance of two-dimensional triangular roughness. J. Hydraulic Research, **15**(1), 19-36.

38. Wijbenga, J., Klaassen, G.J. (1983) Changes in bedform dimensions under unsteady flow conditions in a straight flume. Modern and Ancient Fluvial Systems, ed. by J.D. Collison and J. Lewin, Special Publs. Int. Ass. Sediment, No 6, 35-48.
39. Yalin, M.S. (1964) Geometrical properties of sand waves. J. Hydr. Division, ASCE, **90**(HY5), 105-119.
40. Yalin, M.S. (1977) Mechanics of sediment transport. Pergamon Press, 2nd edition, Oxford, U.K.
41. Yen, B.C. (1991) Channel Flow Resistance: Centennial of Manning's Formula. Water Resources Publications, USA.
42. Znamenskaya, N.S. (1963) Experimental study of the dune movement o sediment. Trans. State Hydrologic Inst. (Trudy GGI), No 108.

4 The Use of Numerical Models in Coastal Hydrodynamics and Morphology

Rolf Deigaard and Jørgen Fredsøe

Department of Hydrodynamics and Water Resources (ISVA), Building 115, Technical University of Denmark, DK-2800 Lyngby, Denmark

4.1 Introduction

Morphodynamics is a complicated field where the interaction between the flow and an erodible bed is studied. Because the bottom changes with time due to erosion or sedimentation, the flow pattern changes correspondingly. This means that a theoretical study of the morphological development of a specific case usually requires not only one flow calculation, but a large number of calculations.

The progress made in morphodynamics during the last decade has been very considerable. A large number of stability analyses have been performed concerning sand waves in rivers, in estuaries and in the sea. Actually a stability analysis can be considered to be a simple morphodynamic problem: you make a perturbation of the bed, calculate the perturbed flow pattern - and investigate whether the perturbation will grow (unstable) or decay (stable) with time. For this purpose, only one flow calculation is needed

A real morphodynamic analysis asks about the further development. This has led to non-linear stability analysis, which can give an answer on some features about the further development in the bed. A major problem with respect to this kind of analysis is that it requires the problem under consideration to be only weakly unstable. In this case the flow can - with some more or less crude approximations - be treated analytically. But the nature is very seldom weakly unstable, so the analysis is often of minor engineering relevance.

One defence for the analytical methods is that you get a clear picture of which parameters are relevant for the problem under consideration, and how they influence the solutions. However you might ask why you should not get the same answer by combining a dimensional-analysis together with a numerical solution. The authors of this paper have worked with Computional Fluid Dynamic (CDF) codes in a number of different topics related to morphodynamics, and our purpose with this paper is to illustrate how much you can gain by applying numerical models in the analysis of morphological development. The emphasis is put on studies the authors have taken part in. The problems considered include free morphological behaviour (instabilities) and the forced development of barred coastal profiles.

4.2 Vortex Ripples

Small scale bed forms formed by waves - sometimes in combination with a current - are very common in a loose seabed. The ripples are usually classified in different types, for a review see Sleath [62], the most common being *rolling grain ripples* and *vortex ripples.*

The rolling grain ripples was named by Bagnold [6]. They are very long, regular bands of rolling grains, which are collected together to form small triangular ridges. Rolling grain ripples are present at very low values of the dimensionless shear stress (Shields parameter, θ), say up to twice the critical Shields parameter.

A more common small scale bed feature is the vortex ripple, where all the sediment in the bed-surface is moving. These exist at larger Shields numbers. In this case we have quite steep bed forms, with a typical steepness (height to length) of 1:5. The length, λ of the ripples are correlated to the amplitude of the near-bed wave orbital motion, a, usually you have $\lambda = 1.3a$. A typical photo of a vortex ripple is shown in Figure 4.1.

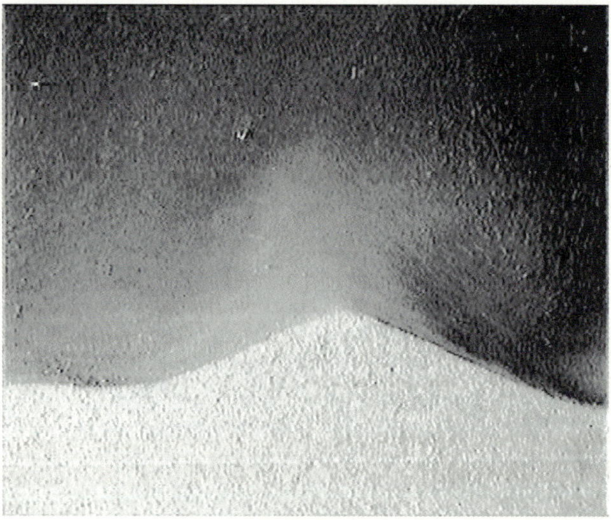

Fig. 4.1. A snapshot of a vortex ripple.

The cause of the formation of ripples can be explained as an instability: you make a small perturbation of the bed, and investigate whether this small disturbance will grow or decay with time. The perturbation is usually sinusoidal in shape, so the instability analysis requires a solution of the oscillatory flow over a wavy wall. The first to do this was Lyne [40], who solved the laminar case for small values of a/λ. This was later extended by Sleath [61] to cover large values of the same parameter. Vittori [68] did the analysis for the relevant magnitude of the parameter $a/\lambda = O(1)$. Foti and Blondeaux [28] further included some turbulent effects in the flow description. For a much more detailed review, see Blondeaux and Vittori [4].

The typical outcome of the flow analysis is the following: the time-averaged (over one period of oscillation) flow gives a streaming pattern, as sketched in Figure 4.2, which is responsible for transporting sediment towards the top of the perturbed bed and thus leads to the instability.

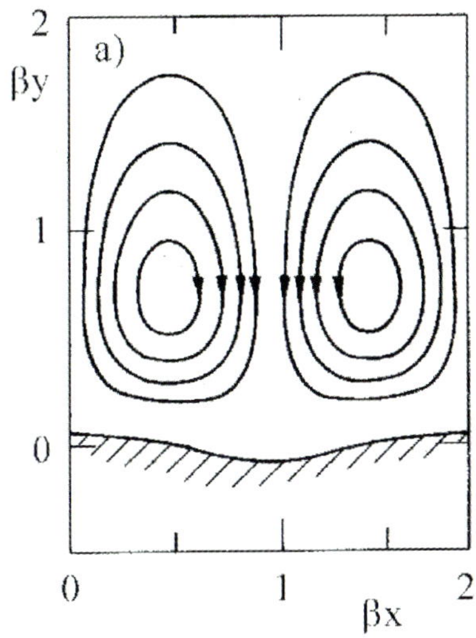

Fig. 4.2. Streamlines of the mean drift over a sinusoidal bed exposed to oscillatory flow, Sleath [62]. β^{-1} is the Stokes length.

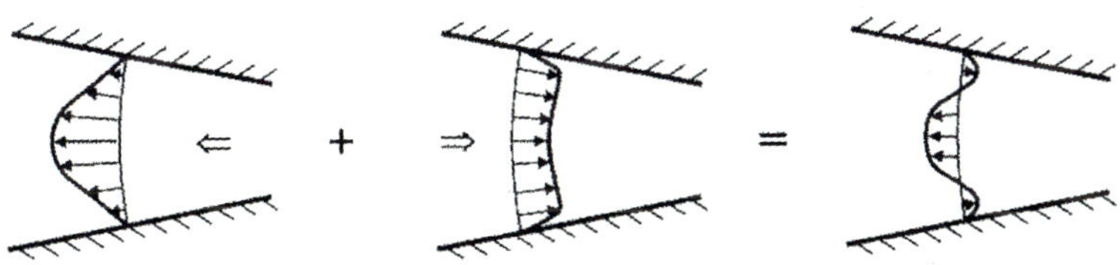

Fig. 4.3. Flow pattern in a divergent flow (left) and convergent flow (middle), which results in streaming pattern as sketched to the right.

The process

The resulting streaming pattern shown in Figure 4.2 is the outcome of the tendency towards flow separation in diverging flows: if we consider a convergent channel as in Figure 4.3, the velocity will be smaller at a certain small distance above the bed when the flow is from right to left (corresponding to the first half wave period) than when the flow is in the opposite direction (second half). So averaged over the whole wave period, there will be a resulting flow component towards right (just above the bed). In a refined flow description, this will lead to a flow pattern like that shown in Figure 4.2.

In a linear perturbation analysis, it is possible to treat this problem analytically/mathematically. Even some nonlinear terms can be included in the analysis as demonstrated by Vittori and Blondeaux [69].

However, many nonlinear terms interact before the vortex ripples are fully developed, and in the final stage, we are no longer talking about tendency towards separation, on the contrary: the flow is fully separated behind the crest of ripple.

This unsteady separating flow can probably not be described using analytical models. A very simple heuristic model was suggested by Fredsøe and Brøker [32]: the main idea was to treat the flow like in Figure 4.3, i.e. splitting the flow up into two half periods, one with flow from left, and another one with flow from right. When the flow started from left, the initial condition was fluid at rest, and from that the flow separation bubble developed with a certain speed which had been determined experimentally. In this way a very simple solution for the ripple shape in its mature stage was developed.

Such a heuristic model can be useful for understanding the processes behind the behaviour of ripples, but quantitatively it has some shortcomings regarding the predictions of wave length and the actual shape of ripples. In such a case it is straightforward to suggest the use of a numerical model to describe the flow and sediment pattern in a proper way.

The numerical modelling

The flow over vortex ripples has been modelled many times in recent years. The favourite model of many researchers is the $k-\epsilon$ model: Sato et al. [59], Aydin [5] and Tsuijmoto et al. [67]. Hansen et al. [33] used as an alternative the discrete vortex model. The latter one also included a description of the suspended sediment due to advection-diffusion.

However only very few scientists go one step further and use the models to do the morphological modelling of the bed ripples. At ISVA we did some attempts to use the discrete vortex model (Hansen et al. [33]), but we never really succeeded: partly because the DVM is very "lively", and partly because the morphological modelling is very difficult.

In a recent Ph.D.-study Ken Andersen together with the second author succeeded in getting some quite promising results (Andersen [2], Andersen and Fredsøe [3]). They used a $k-\omega$ model (a modified version of a $k-\epsilon$ model, which in a more proper way include the no-slip condition at the rough wall and thereby improve the description of the separated flow). The ability of the model to describe the flow is demonstrated in the paper by Fredsøe et al. [31]. The morphological investigation falls in two parts: firstly, the development in the shape of the ripple is investigated, and secondly the dimensions (i.e. length) is treated.

The development in the bed profile is found from the Exner continuity equation

$$\frac{\partial h}{\partial t} = -\frac{\partial q}{\partial x}$$

which states that the bed level change with time is due to gradients in the sediment transport q given as volume including the pore volume. A very accurate discretization of the continuity equation is required to apply this equation to update the bed level during the ripple-growth. Usually a QUICK scheme is being used, Leonard [38]. As demonstrated by Andersen [2], an initial small disturbance of the bed (say a sinusoidal disturbance with arbitrary wave length) will grow with time and attain a final shape, which in the case of dominating contact load will be almost triangular, see Figure 4.4A, with a slope close to the angle of repose, cf. Figure 4.1. This slope is attained because the sediment transport formulation is such, that the sediment transport increases very fast when the angle of repose is approached (avalanches). This very non-linear term is necessary to prevent further growth of the ripples. As also seen from Figure 4.4A, inclusion of suspended sediment makes the ripples less steep. The change is due to the different behaviour of the suspended sediment as compared to the contact load: the suspended sediment transport is not affected by bed slope (directly), and phases are introduced due to the settling and diffusion of suspended sediment. Figure 4.4B demonstrates the robustness of the method: if a current is introduced, the shape becomes asymmetric as expected.

In order to determine the wavelength, Andersen [2] did the following numerical experiment, which is a kind of stability analysis: consider two adjoining ripples with a final developed shape. Now one of the ripples is perturbed by changing its height. Next the exchange of sediment between this and the adjoining ripple decides whether the ripple-pattern are stable or not: if the ripple is made larger, and there will be a net flux (over one wave period) away from this ripple to the adjoining smaller ripples, then the ripple pattern is stable. On the other hand, if the net flux is towards the larger ripple, then the adjoining ripples will be annihilated, and the wavelength will grow. Figure 4.5 shows an example of annihilation of a ripple in a train.

4.3 Coastal Profile Development

Wave induced cross-shore sediment transport can continuously modify a coastal profile, for example by forming new longshore bars, changing the position of existing bars or by making the bars decay. The coastal profile determines the distribution - and to some degree the magnitude - of the longshore wave driven currents and sediment transport.

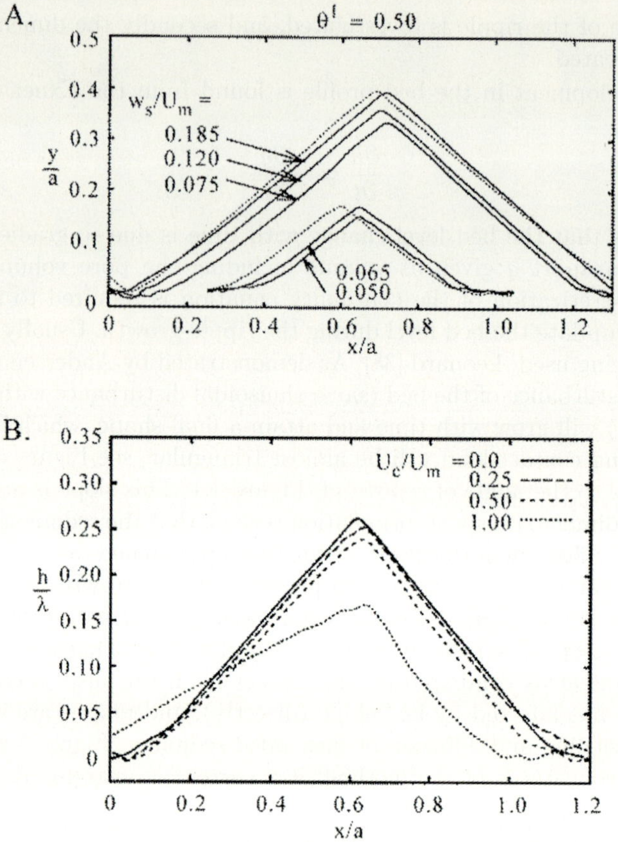

Fig. 4.4. The calculated equilibrium profiles for ripples. A: for waves alone. w_s is the settling velocity, while U_m is the maximum flow velocity. With decreasing values of w_s/U_m the amount of suspended load to bed load increases. B: inclusion of a current.

Determination of the cross-shore sediment transport is difficult because the mean cross-shore discharge is zero, and the sediment transport can be composed of several opposing contributions of the same order of magnitude. Outside the surf zone many contributions have been identified related to non-linear effects in the wave boundary layer, such as: streaming in the boundary layer, drift of suspended sediment, effects of higher harmonics in the orbital wave motion and effects related to wave groups and the associated bound long waves.

In the surf zone the energy dissipation is dominant in the processes governing the wave driven circulation currents and the cross-shore sediment transport. The balance between the radiation stress gradient and the mean pressure gradient, which determines the set-up, is not fulfilled at all levels, even

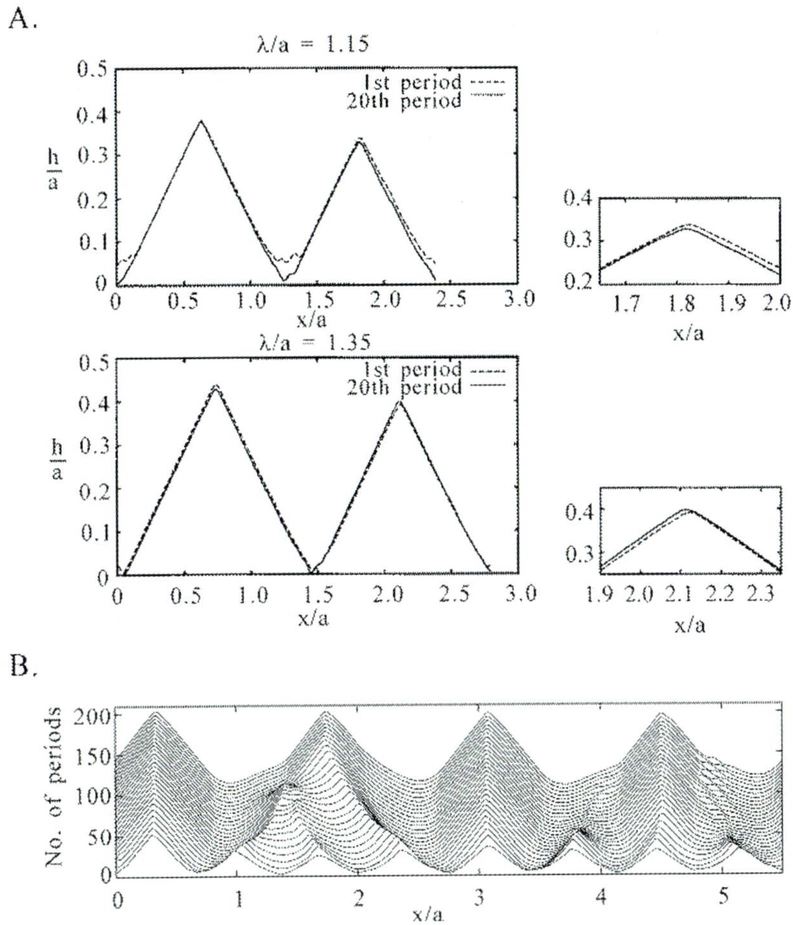

Fig. 4.5. Stability analysis of two ripples. A: the left ripple is higher than the right one, which in the upper figure returns to a higher ripple, i.e. the ripple pair is stable while a longer ripple in the bottom is annihilated after sufficiently long time. B: visualization of annihilation.

though the depth integrated force balance is maintained. This is because the vertical distribution of the two are not the same. In order to obtain momentum balance at all levels, shear stresses must be introduced, and these are associated with a circulation current in the vertical plane, Dyhr-Nielsen and Sørensen [23], Dally and Dean [12] and Svendsen [66]. The shear stress near the surface corresponds to a surface shear stress in the direction of wave propagation and is related to the energy dissipation due to the wave breaking, Deigaard and Fredsøe [18], Deigaard [8]. The circulation current is therefore directed onshore at the surface and - in view of continuity - offshore near the bed (undertow). The sediment transport is also normally found to be in the offshore direction in a surf zone with intensive energy dissipation due to breaking, cf. Figure 4.6.

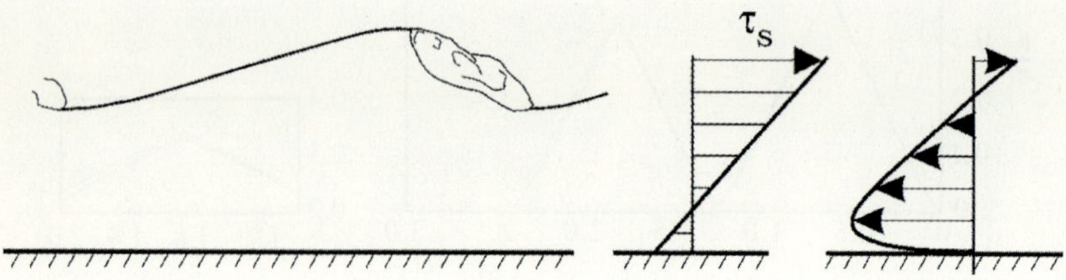

Fig. 4.6. The mean surface shear stress, τ_s, the shear stress distribution and the mean velocity distribution under broken waves in a surf zone.

The offshore directed sediment transport due to the undertow results in a convergence of the transport around the edge of the surf zone which is expected to be the main cause of the formation of longshore breaker bars and the offshore migration of the bars under storm conditions. The sediment transport under non-breaking waves is smaller and is often found to be in the onshore direction mainly due to asymmetry in the wave orbital motion. A long term balance may be possible between the strong offshore transport of short duration under storms and a smaller onshore transport during the longer periods with more gentle wave conditions, Ruessink et al. [57].

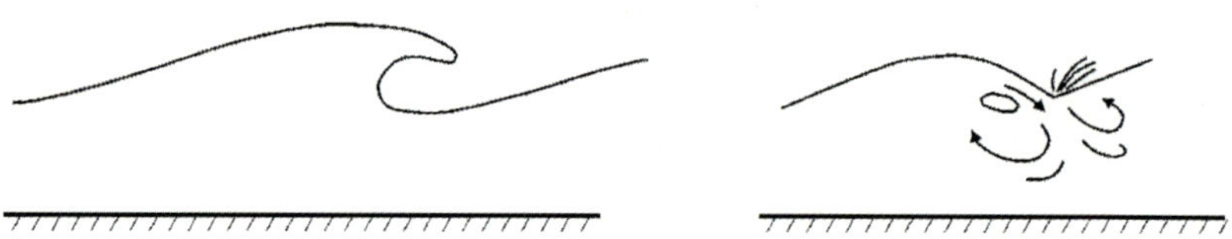

Fig. 4.7. The impact and generation of vortices by a plunging breaker.

Near the point of wave breaking the conditions are complex. The undertow will be established over a distance from the point of breaking, giving non-uniform conditions with a strong circulation current and upward directed-flow at mid-depth. For plunging breakers the breaking takes place as a jet that shoots forward at the wave crest, to plunge down in the trough in front of it, Figure 4.7. The falling jet can create large vortices over the entire water depth (Battjes [7]), that will stir up sediment. It has been proposed that these vortices may play an active role in the formation of the bar under the break point and is important for the geometry of the bar, Miller [19].

Analytical models

The coastal profile has been described in a number of simplified ways, in most cases based on the assumption that there exists an equilibrium profile, and that the actual profile will tend towards this equilibrium.

Dean [13] derived an expression for an equilibrium profile under the assumption that a given bed sediment can only withstand a certain rate of energy dissipation (and thus production of turbulence). If this level is exceeded it will lead to a reshaping of the beach and a wider surf zone with a lower intensity of the wave energy loss. By considering the energy loss per unit volume Dean [13] found the equilibrium profile to be described by the depth, D being proportional to the distance from the coastline x raised to a power 2/3:

$$D \propto x^{2/3} .$$

The Dean profile, which is monotonous without any bars, has been supported by field measurements, and a similar relation was proposed already by Bruun [10] as an empirical relation based on the analysis of beach profiles under different conditions.

Kraus and Larson [35] and Larson and Kraus [37] have developed empirical relations for the development of a barred beach profile, and developed criteria for prediction of the occurrence of beach erosion and beach accretion.

Plant et al. [22] have developed a relatively simple empirical model for the development of a barred profile. The idea is that the profile can be described as a bar that migrates up and down a stationary basis profile. The bar crest is always migrating towards its equilibrium position, which is coincident with the break point of the incoming waves. The rate of the bar response is empirically found to be proportional to the wave height raised to a power three. The predictions of the model has been compared with a long series of surveys at the Coastal Research Facility at Duck, North Carolina.

Numerical models

The analytical models are simplified and to a large extent empirically based. If a detailed description of the development of the coastal profile is required, it must be obtained by a numerical model. Numerical models are typically composed of a wave module, a module for the wave driven circulation current, a sediment transport model and a module for the morphological change of the profile. The sequence of modules is then run for each morphological time step.

Some of the first models based on this principle were presented by Dally and Dean [12] and Stive and Battjes [65]. An extensive review of the development of numerical profile models, the strategy behind the models and the different processes included in them is given by Roelvink and Brøker [56], who also present comparisons between the predictions by a number of selected models. One of the models described, Pechon [48], is a two-dimensional vertical model that includes some of the effects of the non-uniformity in the wave driven circulation current near the break point. In the following results are given from two numerical models that address specific aspects of cross-shore sediment transport and profile modelling.

The first model, Pedersen et al. [50], simulated the sediment transport due to plunging breakers near the break point. The model is two-dimensional in the vertical plane with a movable linearized surface boundary. A wave with decaying height is imposed as a surface boundary. The hydrodynamic model is a discrete vortex model using the cloud-in-cell method. The plunging wave is simulated by introducing a jet of water that falls down and penetrates the surface in front of the wave crest. [29]. A corresponding amount of water is re-

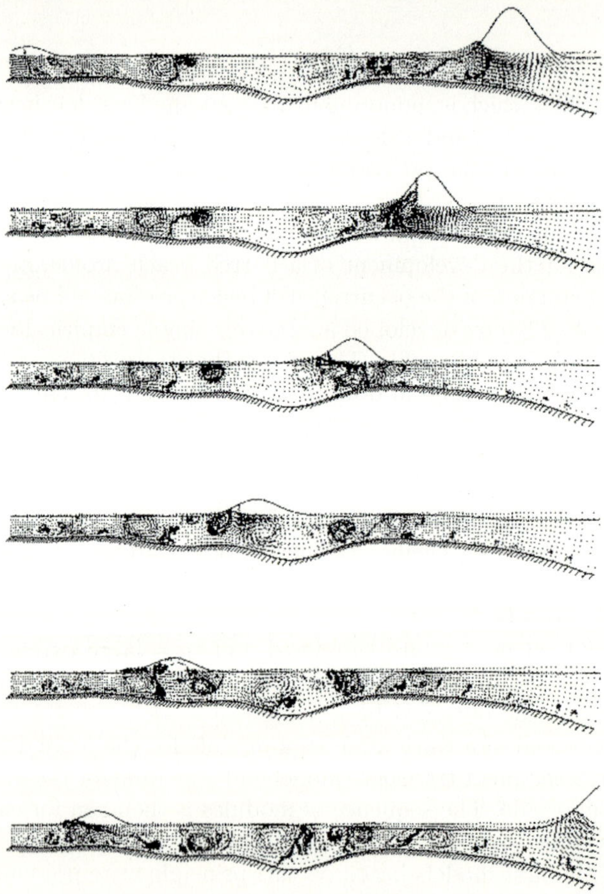

Fig. 4.8. Simulated velocity and suspended sediment field in a plunging breaker, Pedersen et al. [50].

moved from the wave crest. Near the bed a turbulent boundary layer model is used, Fredsøe The sediment transport is described as bed load and suspended load transport. The bed boundary condition for the suspended sediment is given as a reference concentration c_b at a distance of two grain diametres from the bed. The reference concentration c_b is a function of the instantaneous bed shear stress, Engelund and Fredsøe [11]. In the near bed turbulent boundary layer the suspended sediment field is described by the turbulent diffusion counteracting the settling. Away from the bed the suspended sediment is described by a Lagrangian simulation, advecting sediment 'particles' with the flow field induced by the wave motion and the plunging process.

Figure 4.8 shows the flow field and clouds of the suspended particles at different phases during a wave period. It is seen that the jet simulating the plunging creates large vortices that can entrain the suspended sediment over the entire water depth. In Figure 4.9 an example of an instantaneous suspended sediment concentration profile near the plunge point is given. It can be seen how the model can produce inverted concentration profiles with the highest sediment concentration away from the bed. The time-averaged

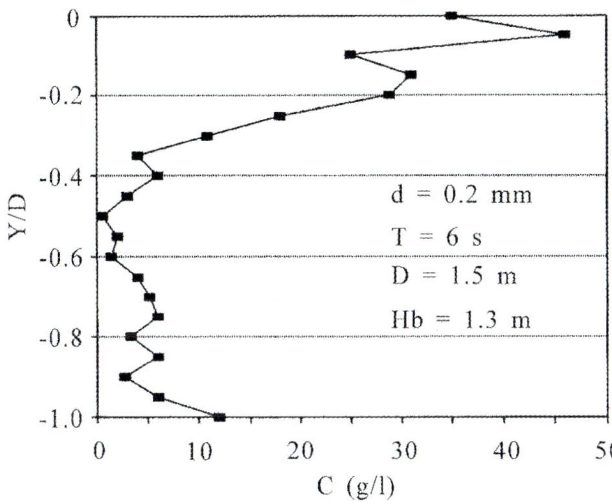

Fig. 4.9. Example of instantaneous concentration profile close to the plunge point.

concentrations will, however, decrease towards the surface as seen in Figure 4.10, where the time mean concentrations near the plunge point are shown for different grain sizes. For finer sediment the concentration is more evenly distributed over the vertical than for the coarser sediment, but the profile is not as sensitive to the settling velocity here, where the concentration field is mainly determined by advection, as when the concentration distribution is determined by turbulent diffusion and settling.

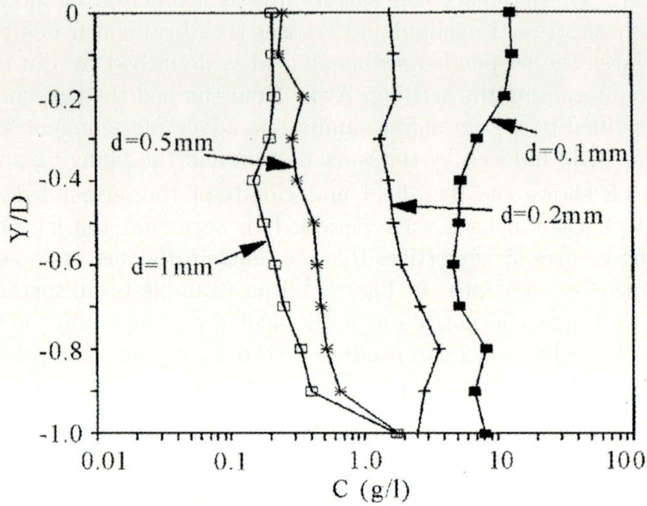

Fig. 4.10. Time-averaged concentration profiles for four different grain sizes of sediment at a position close to the plunge point.

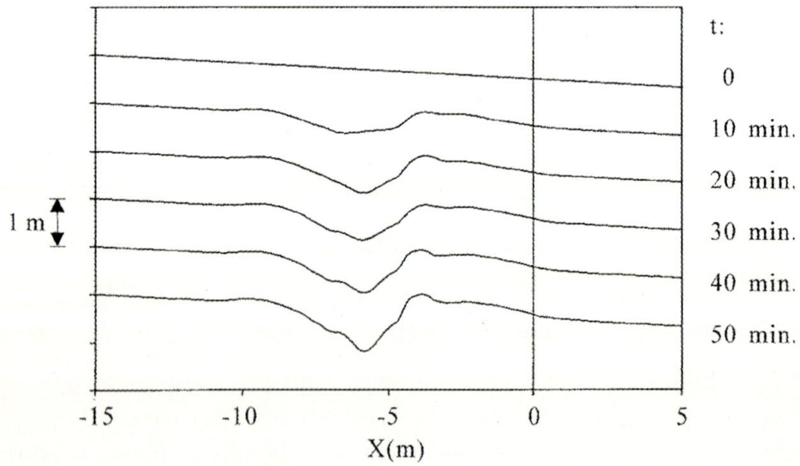

Fig. 4.11. Simulated morphological development of a coastal profile under plunging breakers, Pedersen [49]. Data: $T = 6$ s, $H_b = 1.3$ m, $d = 0.2$ mm.

A morphological simulation is presented in Figure 4.11 where it is seen that the present simulations can create a bar and a trough inshore of it. The model demonstrates that the effect of plunging may be simulated, that plunging breakers can stir up considerable amounts of sediment over the vertical and that this may play a role in the formation of breaker bars. The combined effect of plunging breakers and the wave driven circulation current (undertow) remains to be investigated.

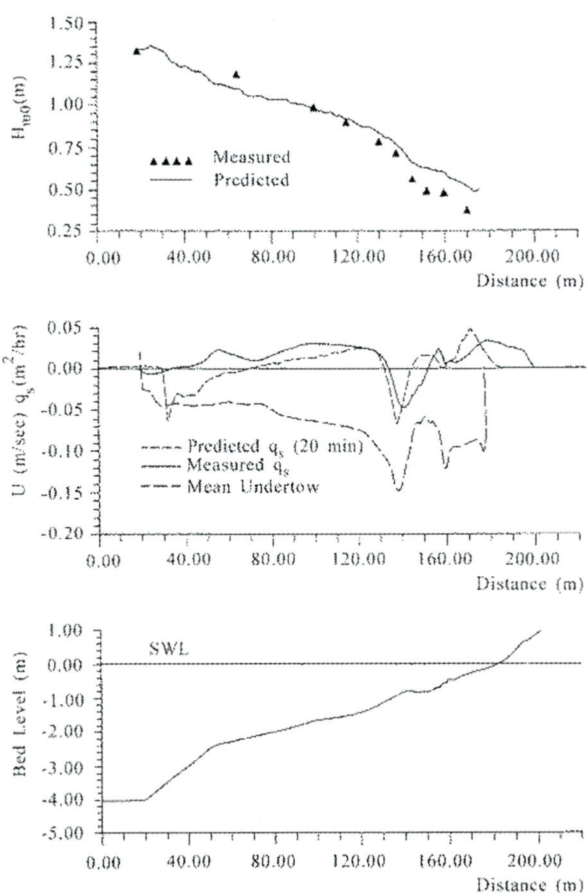

Fig. 4.12. Comparison between measured cross-shore sediment transport rate in test 1b of the LIP 11D experiment and simulations by Rakha et al. [23]. Top: variation in wave height across profile. Middle: measured cross-shore sediment transport rate (full drawn curve), simulated cross-shore sediment transport rate and simulated depth averaged undertow velocity. Bottom: coastal profile, measured.

The second study has been based on a detailed phase resolving wave modelling, using a Boussinesq wave model for surf zone conditions. One of the promising methods to simulate breaking and broken waves in a surf zone by Boussinesq models is by use of the surface roller concept, Deigaard [7], Schäffer et al. [60], Madsen et al. [41,42]. Rakha et al. [23] made a model for cross-shore sediment transport and profile development on the basis of the Boussinesq wave model of Madsen et al. [41], [42]. The Boussinesq wave model gives time series of the instantaneous near bed wave orbital velocity, the water depth and the dissipation due to breaking in all points of the computational grid across the profile. Irregular waves are modelled, and the low frequency wave motion is also represented by the model. The results from the Boussinesq wave model are used to drive a boundary layer model, Fredsøe [29], a model for the wave generated turbulence, Deigaard et al. [19], a model for the wave generated circulation current and a sediment transport model, Fresdøe et al. [30].

The Boussinesq model predicts the horizontal asymmetry in the wave induced horizontal velocities and the coupling between the low and the high frequency motions. The model can reproduce the onshore transport direction, that is primarily caused by the asymmetry in the wave motion, which is dominant outside the surf zone, and the offshore directed transport due to the stronger undertow in the areas with the strongest energy dissipation due to wave breaking. Figure 4.12 shows an example of a comparison between the calculated cross- shore transport and measurements in the Delta Flume from one of the LIP11D experiments carried out at Delft Hydraulics, Arcilla et al. [4].

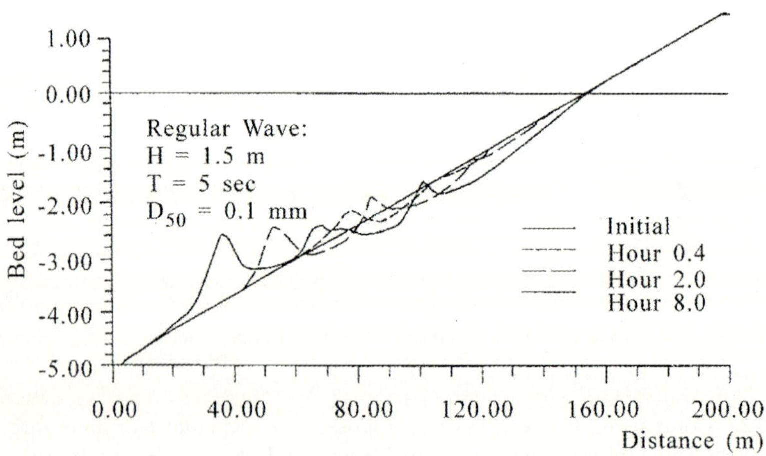

Fig. 4.13. Simulation of coastal profile development, Rakha et al. [23].

Figure 4.13 gives an example of a morphological profile simulation with the model of Rakha et al. [23]. It is seen how bars are formed and migrate offshore while they grow in volume but with a decreasing crest level.

Profile models have been compared with large scale laboratory experiments in wave flumes, as for example the Big German Wave Channel (Dette and Ulizcka [20]), the Delta Flume of Delft Hydraulics (Arcilla et al. [4]) and the SUPERTANK experiments (Kraus et al. [36]). There is agreement between the experimental conditions for these tests and the assumptions behind the profile models. It has not - without a very specific calibration - been possible for detailed process-based profile models to describe the evolution over longer periods with a sequence of erosive and accretive events. The profile models have primarily focussed on the erosive events. It is not at present known to which degree the purely two-dimensional processes, that are considered in the profile models, are actually responsible for the morphology of natural coasts, and if even weak three-dimensional flow phenomena are of importance, such as a horizontal circulation over a weakly non-uniform coastal bathymetry.

4.4 Shear Waves

Observation of the phenomenon of shear waves was first reported by Oltman-Shay et al. [44] in an analysis of measurements under the SUPERDUCK experiments at the field research facility at Ducks, North Carolina. Fluctuations in the longshore current were demonstrated to be associated with waves propagating along the coast. The shear waves are always found in connection with a longshore current, propagating in the current direction with a celerity close to half of the maximum of the time averaged longshore current. The propagation velocity is thus independent of the wave length of the shear waves. The wave lengths are so short compared to the frequency of the shear waves that they could not be explained by any theory for surface gravity waves; initially the waves were therefore termed 'Far Infra-Gravity Waves'.

The occurrence of the shear waves was explained in a companion paper, Bowen and Holman [9], as being the result of an instability of an otherwise steady, uniform wave-driven longshore current, analogous to the instability of a free shear layer flow. A uniform wave-driven longshore current was assumed to be given a small perturbation that is periodic in the longshore direction. It is then investigated if the perturbation tends to grow in time (in which case the uniform longshore current is unstable and cannot exist without developing shear waves) or if it tends to decay in time (in which case the longshore current is stable with respect to the particular perturbation considered). The unstable perturbation with the most rapid exponential growth is often interpreted to some degree to be representative of the actual non-uniform and unsteady flow that will emerge. The linear stability analysis of Bowen and

Holman [9] is very simple, but its results have in essence been confirmed by later more sophisticated calculations.

Although the occurrence of shear waves is expected to be quite frequent they have only been demonstrated to be present in rather few cases, e.g. Oltman-Shay and Howd [45]. A single comprehensive laboratory experiment has been conducted, Reniers and Battjes [54] and Reniers et al. [55]. In these experiments one of the findings was that shear waves readily developed with a wave-driven longshore current concentrated over a longshore bar, while a longshore current along a monotonous beach profile could remain stable. An extensive review of the last ten years of research on shear waves is given by Dodd et al. [21].

Analytical models

The linear stability analysis of Bowen and Holman [9] has been refined in many ways to make the conditions treated more realistically. The most important refinements can be summarized as follows:

- Inclusion of bed shear stress in the flow description, Dodd and Thornton [22]. The inclusion of bed friction is found to make the longshore current less unstable. The bed friction has both an effect on the predicted initial longshore current (the zero order solution) and on the character of the perturbations, which may turn out to be stable (decaying in time) or unstable (growing in time).
- Momentum exchange in the horizontal was included by Falqués and Iranzo [25] as an eddy viscosity. The momentum exchange has influence on the initial longshore current distribution as well as on the flow perturbations. The momentum exchange was found to act stabilizing on the longshore current.
- The significance of the shape of the coastal profile was investigated by Putrevu and Svendsen [52], who found that a longshore bar could have a significant destabilizing effect on the longshore current. The stronger tendency for instability can be because the flow on a bar with a trough inshore is less restricted compared to a monotonously sloping profile towards the shoreline. Another reason may be that the longshore velocity profile is narrower over the bar, with an inflection point on either side of the location of the velocity maximum, Figure 4.14. Putrevu and Svendsen [52] used velocity profiles derived for a constant sloping beach and have thus concentrated on the first effect.

A linear stability analysis as those described above can give important information about a longshore current system. Above all, it can predict if a steady uniform longshore current can exist, or if shear waves will develop. It does, however, assume that the perturbations are infinitely small and can therefore not give information on the strength of the shear waves or on their effect on for example the mean longshore current distribution, the sediment transport or the dispersion of matter in the nearshore zone. The unstable

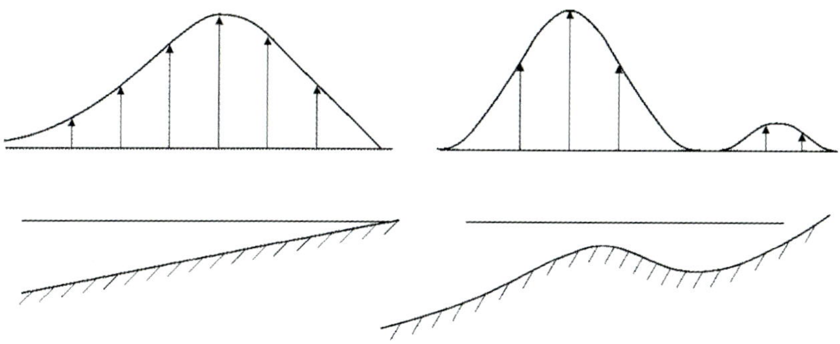

Fig. 4.14. The longshore current over a constant slope beach and over a uniform barred beach profile.

perturbations leading to shear waves develop relatively rapidly and in natural conditions the shear waves will practically always be fully developed.

Dodd and Thornton [22] extended the analysis to a weakly non-linear situation, taking advantage of the fact that the instability only occurs if the coefficient describing the bed friction (as an example) is smaller than a certain critical value. When the friction is close to critical, the strength of the shear waves described by the most unstable linear perturbation can be determined as function of the deviation from the critical value.

Numerical modelling

If a description of fully developed shear waves is wanted, the unsteady flow becomes very complex including large scale irregular eddies, so it can in general only be obtained by a numerical model that solves the flow equations for the wave driven currents in the domain considered. At present all simulations of shear waves have been performed by two-dimensional depth integrated models.

Falqués et al. [26] made a numerical modelling study of fully developed non-linear shear waves on a constant slope beach, studying among others the transfer of energy between the most unstable (linear) perturbation and the sub- and super-harmonics, as the strength of the perturbation increases.

Deigaard et al. [9] made a study by applying the numerical modelling system MIKE21 of DHI. They studied shear waves and the sediment transport over a coastal profile with a longshore bar. In the following some examples of the results are given. The coastal profile is shown in Figure 4.15 together with the cross-shore variation of the wave height. The incoming waves have a significant wave height of 3.0 m, a period of 9 s and a direction of 45 degrees relative to the shore normal, all taken at the offshore boundary at x equal zero. The simulation is made over a domain with a length of 3600 m.

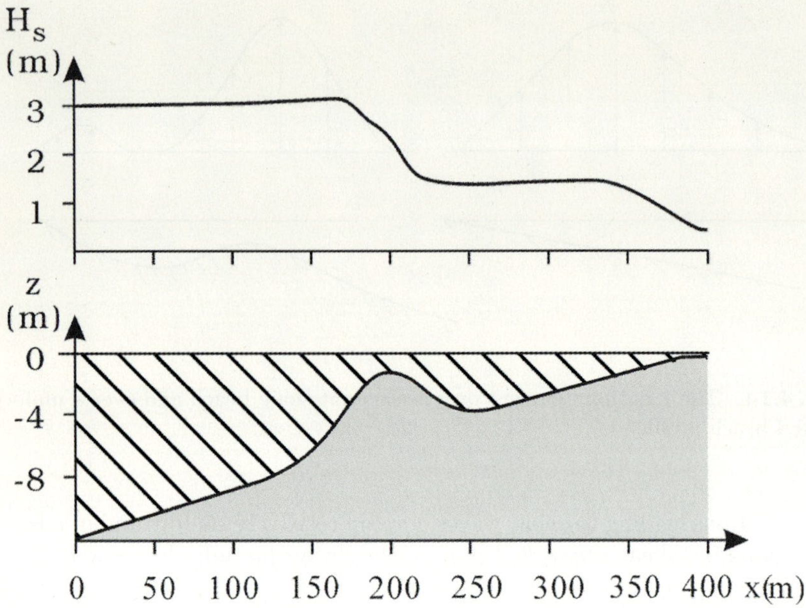

Fig. 4.15. Bottom: coastal profile. Top: simulated wave height variation across the profile.

Figure 4.16A shows vector plots of the velocity field in the downstream part of the simulation domain. The driving force is gradually increased from zero over the first 2000 s to become stationary thereafter. It may be noted that the frequency/wave number of the shear waves is reduced during the time elapsed from the first to the second vector plot in Figure 4.16A. This phenomenon is treated in detail by Allen et al. [2] and Slinn et al. [63]. Figure 4.16B shows the instantaneous water surface elevation along the bar crest for the fully developed shear waves. The spatial development of the perturbation can be seen, the shear waves become visible at some distance from the upstream boundary, they grow in amplitude to reach an approximately constant level. The wave length is about 190 m and the phase speed is 0.95 m/s, 55% of the maximum longshore current velocity.

The shear waves cause a redistribution of the momentum across the profile analogous to the effect of Reynolds' stresses in a turbulent flow. Figure 4.17 shows the redistribution of the longshore momentum due to the fully developed shear waves. Figure 4.17A gives the distribution of the three elements in the time-averaged longshore momentum balance:

$$-\frac{\partial S_{xy}}{\partial x} = \overline{\tau_b} + \frac{\partial}{\partial x}\overline{(\rho D u v)}$$

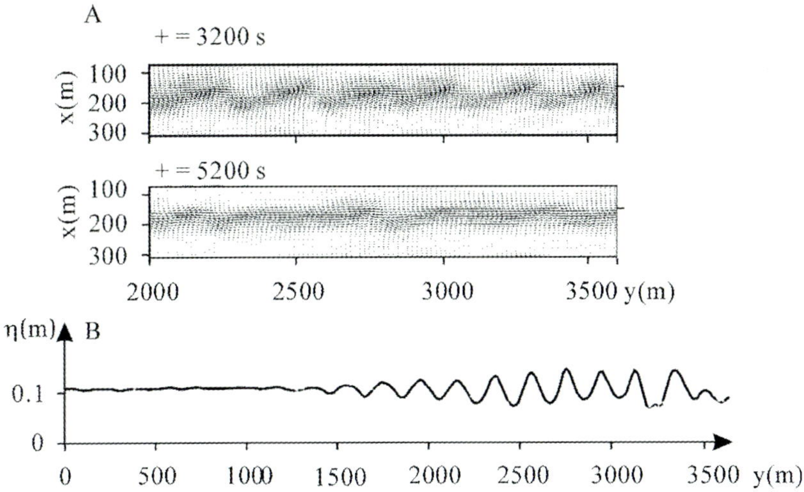

Fig. 4.16. A: vector plots of the velocity field at the downstream part of the bar. B: water surface elevation along the bar crest, fully developed shear waves.

where S_{xy} is the shear component of the radiation stress, τ_b is the bed shear stress, ρ is the density of water, D is the depth and u, v are the velocity components in the x- and y-direction. An overbar signifies a time average over a time scale large compared to the shear wave period.

Figure 4.17B shows the mean longshore current and, for comparison, the longshore current corresponding to a balance between the driving force and the bed shear stress in the absence of any shear waves. The shear waves cause a considerable smoothing of the velocity distribution and a reduction of the maximum velocity, but the location of the maximum velocity has not changed. The mechanism of the momentum exchange is illustrated in Figure 4.18 which shows traces of the velocity vectors at three locations: at the bar crest and 40 m offshore and inshore of the crest. The correlation between the longshore and cross-shore velocity fluctuations clearly gives a non-zero cross-shore flux of momentum at the bar crest and offshore of it, while at the inshore position the correlation is small, and the momentum exchange is equally small even though the magnitude of the velocity fluctuations is considerable.

By varying the flow resistance (described by the Manning number) and the momentum exchange coefficient (described as an eddy viscosity) it was found that an increase of the bed friction or of the momentum exchange can suppress the formation of shear waves in the model.

Figure 4.19 illustrates the effectiveness of the shear waves in the dispersion of matter by showing the instantaneous plume of a substance released at the bar crest at the upstream boundary. The simulation is made with a specified

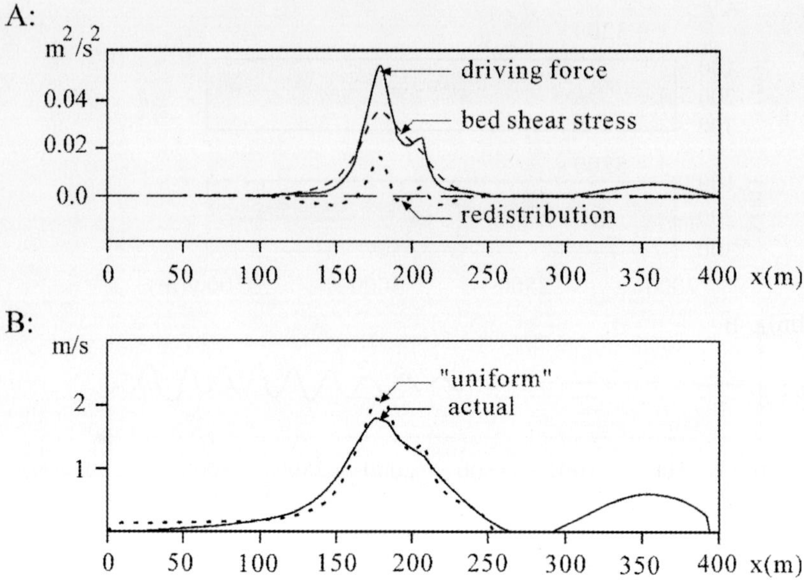

Fig. 4.17. A: full drawn line: $\partial(S_{xy}/\rho)/\partial x$, broken line: $\bar{\tau}_b$, dotted line: $\partial(\overline{Duv})/\partial x$. B: Time mean longshore velocity, full drawn curve: actual velocity; dotted curve: assuming steady, uniform conditions.

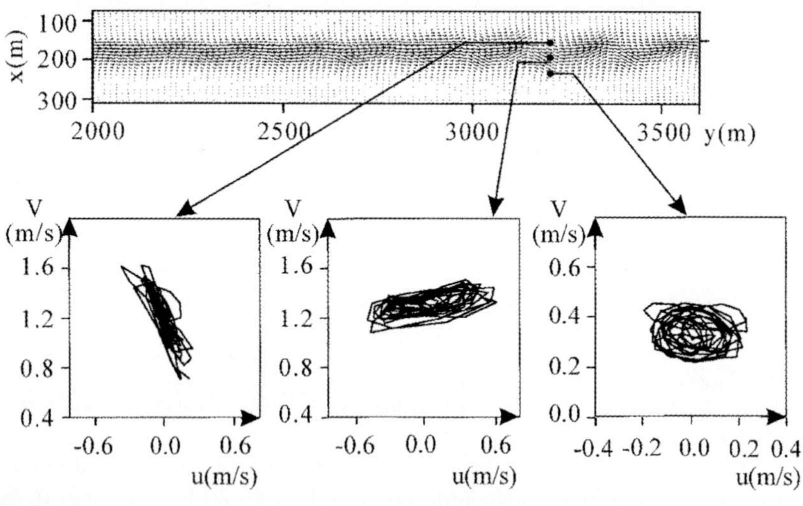

Fig. 4.18. "Phase plots" showing the traces of the current velocity vectors at three locations; v is the longshore and u is the cross-shore velocity.

turbulent exchange coefficient of zero. The shear waves may be described as two rows of vortices that are advected with the longshore current. The two rows are on each side of the maximum in the mean longshore current, and the vortices in the two rows rotate in opposite directions. It can be seen that the plume is deformed by the shear waves and is rolled up to form unconnected 'cat's eyes' around each vortex in the inshore row of vortices.

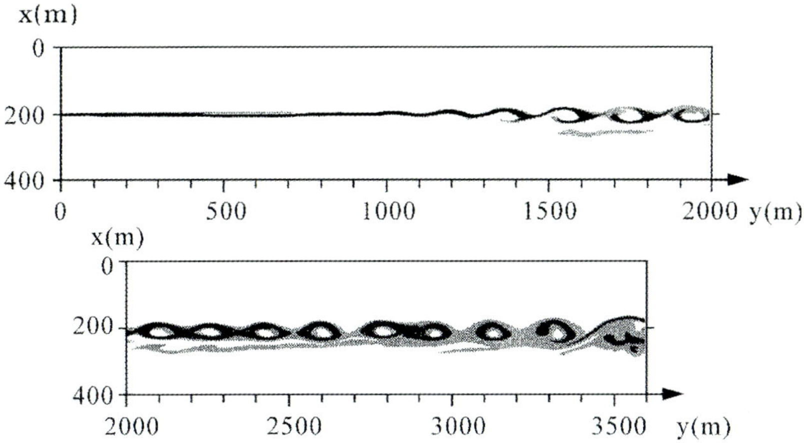

Fig. 4.19. Dispersion of dissolved material discharged at the bar crests.

In a numerical model it is simple to deviate from the assumption of uniform conditions along the coast. Deigaard et al. [9] made simulations with rhythmic alongshore variations in the crest level of the longshore bar and found that this was quite effective in suppressing the formation of shear waves, causing instead a stationary wave pattern in the flow, mainly due to the variation in the forcing caused by the alongshore variation in the crest level.

The shear waves over the alongshore uniform bed topography had a noticeable effect on the sediment transport. The magnitude and distribution of the time-mean longshore transport is modified as seen in Fig. 4.20. The shear waves reduced the calculated longshore sediment transport rate by 17%. Due to the non-linearity of the sediment transport relations the cross-shore velocity fluctuations were found to cause a mean cross-shore sediment transport which is not negligible compared to contributions from other cross-shore transport contributions, Fig. 4.20B.

Numerical models have been used to investigate many aspects of the hydrodynamics of shear waves. Özkan and Kirby [46] and Özkan-Haller and Kirby [47] have simulated the conditions in the SUPERDUCK field experiment. Allen et al. [2] have investigated the conditions over a constant slope

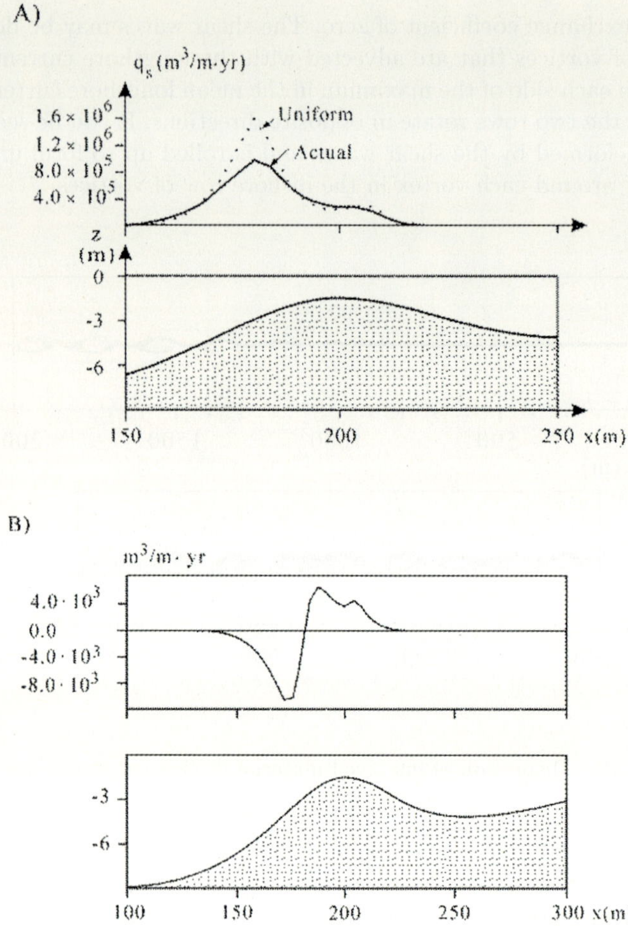

Fig. 4.20. A: the distribution of the time mean longshore sediment transport across the bar profile. B: the distribution of the time-mean chross-shore sediment transport.

beach profile, they analysed the influence of the length of the simulation domain, and the transfer of energy to subharmonics as the basic situation considered departs more and more from the critical condition. Slinn et al. [63] considered barred as well as constant slope beach profiles. They made simulations of different shear wave intensity and classified different flow regimes with turbulent shear flow as the highest. Sancho and Svendsen [29] made simulations of the wave driven current over a non-uniform barred profile with a rip channel cutting through the bar, investigating the effect of the rip channel on the formation and the characteristics of shear waves in its vicinity.

4.5 Coastal Stability

It is often observed that the bathymetry along a long straight coast is not uniform, but exhibits a more or less regular pattern of bed features, typically with a length scale of several times the width of the surf zone as described by for example Sonu [64] and Lippman and Holman [39]. The explanation for the non- uniformities can be that a long straight uniform coast is in fact unstable under certain boundary conditions: height and direction of incoming waves, beach profile, grain size and sediment transport regime, and the instability can be treated by a linear stability analysis.

In the analysis of the coastal stability the uniform coast is assumed to be given an infinitely small perturbation that is varying harmonically in the alongshore direction. The bed perturbation will cause a small change in the wave field, the driving forces from the waves (radiation stresses) and in the wave driven currents. The change in the wave and the current field will modify the sediment transport pattern and give a field of erosion and deposition. The erosion/deposition field will change the bed perturbation and can make it grow or decay in addition to a possible migration along the coast. If the perturbation is decaying the uniform coast is stable with respect to this particular perturbation. The possible bed perturbations that are amplified (or diminished) without changing their basic shape are determined as solutions to an eigenvalue problem defined by the equations for the hydrodynamics, the sediment transport and the bed level change. In the linear stability analysis the perturbations are assumed to be infinitely small, but the fastest growing perturbation is often assumed to be representative for the characteristics of the finite bed features that emerge under the given conditions. Although the physical processes are completely different, the mathematical treatment of the coastal instability is in many respects analogous to the analysis of the instabilities leading to shear waves.

When the waves approach the coast at an angle and drive a longshore current there is a high degree of similarity between the mechanisms of the coastal instability, with an interaction between the longshore current and the bed perturbation, and the instability of a river forming alternate bars as the first stages of meandering. The main differences between the two cases are in the geometry with the river flow restricted by the banks and the coast, which is unbounded to the offshore but has a localized longshore driving force in the surf zone. Other important differences are due to the effects of the waves. The coastal bed perturbation modifies the pattern of wave breaking and thus the forces driving the current, for example by giving an additional shoreward forcing where the perturbation makes the depth in a surf zone shallower. The combined effect of waves and current on the sediment transport is also significant. If the flow along a streamline goes in the direction of decreasing wave height (for example towards a more shallow area in a surf zone) this may cause deposition of sediment, even if the current velocity is increasing along the streamline.

eigensolution had been bound: the bed perturbation migrates and grows (or decays) without changing its shape.

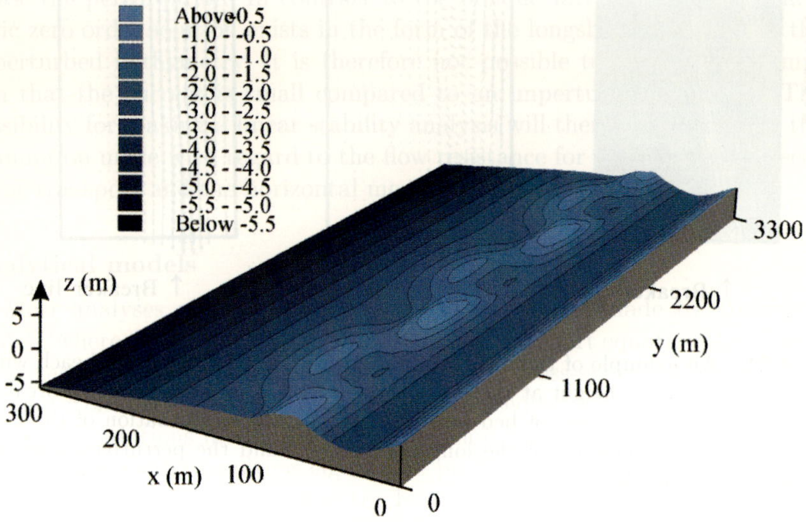

Fig. 4.22. An example of the most unstable perturbation superposed on the coastal profile.

Figure 4.22 gives an example of the most unstable perturbation on a coast with a depth D_c of 1.0 m over the bar crest. The alongshore wave length of the perturbation is 1100 m. The deep water significant wave height is 1.1 m with an angle of 20 degree form normal wave incidence. An arbitrary magnitude of the perturbation has been chosen to illustrate the bathymetry which the instability tends to form. It is seen that the perturbation forms a meander-like pattern over the bar forming a series of rip channels with a berm offshore of each rip channel. On the bar between the rip channels the perturbation tends to steepen the offshore face of the bar. Although there is nothing in the present analysis to prevent the formation of oblique bed forms, there is practically no tendency for the instability to be oblique, in contrast to what has been found in stability analyses for constant sloping beaches (cf. Hino [34], Christensen et al. [11] and Falqués et al. [27]). The coastal profile is shown in Figure 4.23 with the perturbation from the locations of the rip channels and from the locations between the rip channels added. The cross-shore integral of the perturbation is close to zero, and the instability is mainly formed by shifting sediment from the bar crest towards the offshore at the location of the rip channels and from offshore to the bar crest between the rip channels.

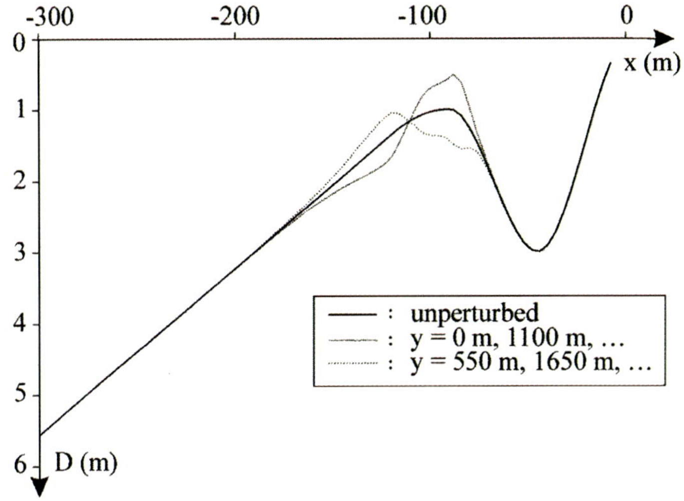

Fig. 4.23. The coastal profile with the unstable perturbation superposed at the position of the rip channel, and at the position between the rip channels.

By analysing the flow, sediment transport and the erosion/deposition, it is found that the growth of the instability is closely related to the cross-shore flux of water over the bar crest. The cross-shore flow component gives gradients in the wave conditions (wave height, energy dissipation rate) along a streamline due to the variation across the profile. An onshore flow first causes erosion on the offshore part of the bar, with a flow into the surf zone with increasing sediment transport capacity, and then deposition in the saturated surf zone where all waves are breaking and the wave height decreases in the flow direction.

The analysis has been performed for a range of alongshore wave lengths, maintaining all other parameters constant. Figure 4.24 shows the amplification ω_i and the migration velocity a as function of the wave length L_p. It is seen that the amplification has a very clear maximum at a wave length close to the 1100 m, which is then characterizing the most unstable perturbation and the expected preferred wave length of a rip channel system. The exponential growth of the instability corresponds to a doubling in about 8 hours, and the migration velocity of the fastest growing perturbation is about 80 m/day going in the longshore transport direction.

The significance of different parameters was investigated by the models, and it was found that the most unstable alongshore wave length depends on the geometry of the trough inshore of the bar, this is because of the flow in the trough generated by the shore-normal flow over the bar crest. The

56. Roelvink, J. A. and Brøker, I. (1993) Cross-shore profile models. Coastal Engineering, Vol. 21, pp. 163-191.
57. Ruessink, B. G., Houwman, K. T. and Hoekstra, P. (1999) Medium-term frequency distributions of cross-shore suspended sediment transport rates in water depths of 3 to 9 m. Coastal Engineering, Vol. 38, pp. 25-46.
58. Sancho, F. E. and Svendsen, I. A. (1998) Shear waves over longshore nonuniform barred beaches. Proc. 26th Int. Conf. on Coastal Engineering, ASCE, Copenhagen 1998, pp. 230-243.
59. Sato, S., Uehara, H. and Watanabe, A. (1986) Numerical simulation of the oscillatory boundary layer flow over ripples by a k- turbulence model. Coastal Engineering, Japan, Vol. 29, pp: 65-78.
60. Schäffer, H. A., Madsen, P. A. and Deigaard, R. (1993) A Boussinesq model for waves breaking in shallow water. Coastal Engineering, Vol. 20, pp. 185-202
61. Sleath, J.F.A. (1976) On rolling grain ripples. J. Hydraulic Res., Vol. 14, pp: 69-81.
62. Sleath, J.F.A. (1984) Sea bed Mechanics, Wiley.
63. Slinn, D. N., Allen, J. S., Newberger, P. A. and Holman, R. A. (1998) Nonlinear shear instabilities of alongshore currents over barred beaches. J. Geophys. Res., Vol. 103, pp. 18357-18379.
64. Sonu, C.J. (1972) Field observations of nearshore circulation and meandering currents. J. Geophys Res., Vol. 77, No. 18, pp. 3232-3247.
65. Stive, M. J. F. and Battjes, J. A. (1984) A model for offshore transport. Proc. 19th Int. Conf. on Coastal Engineering, ASCE, pp. 1420-1436.
66. Svendsen, I. A. (1984) Mass flux and undertow in a surf zone. Coastal Engineering, Vol. 8, pp. 347-365.
67. Tsujimoto, G., Hayakawa, N., Ichiyama, M., Fukushima, Y. and Nakamura, Y. (1991) A study on suspended sediment concentration and sediment transport mechanism over rippled sand bed using a turbulence model. Coastal Engineering in Japan, Vol. 34(2), pp: 177-189.
68. Vittori, G. (1989) Nonlinear viscous oscillatory flow over a small amplitude wavy wall. J. Hydraul. Res., Vol. 27, pp. 267-280.
69. Vittori, G. and Blondeaux, P. (1990) Sand ripples under sea waves. Part 2. Finite amplitude development. J. Fluid Mech., Vol. 218, pp. 19-39.

5 Process of Occurrence, Flow and Deposition of Viscous Debris Flow

Tamotsu Takahashi

Disaster Prevention Research Institute, Kyoto University Gokasho, Uji, Kyoto 611-0011, Japan

5.1 Abstract

Debris flow generally contains particles whose size distributes in a very wide range, and the fine fraction of them is usually incorporated in the interstitial water to behave as a kind of fluid. When the fine fraction is composed of cohesive solids and it accounts for a significant ratio in the entire particle fractions, the interstitial fluid becomes highly viscous slurry. In addition to the large viscosity of the slurry, the dense content of coarse particles enhances the apparent viscosity of the whole material. Therefore, in such a debris flow, the viscous stress is more dominant than the stresses due to the inter-particle collisions and/or the macro turbulent mixing. This kind of debris flow is called as the viscous type which typically occurs at Jiangjia Gully, Yunnan, China. Field observations of running viscous debris flows and some field inspections of the debris flow deposit suggest that viscous debris flow arises from landslide. A qualitative model for the process of transformation of a soil mass into debris flow is introduced. Once debris flow is generated it surges down the slope or the channel in fantastic speed. Such a small resistance to flow can only be fulfilled by nearly homogeneous dispersion of coarse particles within flow. Here, the particle dispersion mechanism in laminar flow is attributed to both the drag of the squeezing flow in the viscous interstitial fluid and the higher resistance to the particle motion toward the more densely particle concentrated area. When a viscous debris flow comes down to a mildly sloping area its competence for dispersing the coarse particles diminishes due to decrease in the velocity of flow, and it begins to deposit. The sediment concentration in viscous debris flow is so dense that the volume change before and after deposition is little, and the fresh deposit is very soft, virtually not different from the material in flow. A newly arriving surge easily entrains the deposit and goes together downstream. Soon, as the depth of flow and the velocity decrease in the rear part, deposition takes place again. Repetition of such the phenomena continues long at Jiangjia Gully. A Newtonian fluid model introduced herein is examined by the flume experiments as well as by the field observations.

5.2 Introduction

Debris flows can be defined as highly concentrated flow of the mixture of water and solids that run down steep slopes or ravines in response to gravitational attraction. Although the particle sizes distribute widely in general, the composition is different case by case reflecting mainly the geology of the area. Depending on the particle composition, the solids concentration, the thickness of flow, the slope gradient down which the flow surges, etc., debris flows appear in different behavior. When the material contains virtually no fine fraction such as clay and silt, if the relative depth of flow; i.e., the ratio of the depth of flow, h, divided by the mean diameter of the coarse particles contained in the flow, d, is less than about 30 and the solid concentration, C, is larger than about 0.2, the stony type debris flow appears, within which the coarse particles move colliding and jostling to concentrate the largest particles at the forefront of the surge, and if the relative depth is larger than about 30, the turbulent muddy type debris flow appears, within which particles are suspended and violently mixed by the turbulence in flow Takahashi [9]. When the material contains significant amount of fine cohesive particles, the interstitial fluid becomes highly viscous slurry, and the viscous type debris flow appears, of which characteristics are discussed in detail later in this paper.

Mechanically, the difference in behavior of debris flows must be the reflection of the difference in dominant stresses among others existing in flow. If the total shearing stress, τ, and the pressure, p, in a very highly sediment-laden plain shear flow are given by the linear sum of the constituent shearing stresses and pressures, they will be described as following:

$$\tau = \tau_y + \tau_\mu + \tau_c + \tau_k + \tau_s + \tau_t \tag{5.1}$$

$$p = p_c + p_s + p_h \tag{5.2}$$

where τ_y is the yield strength of the interstitial fluid, τ_μ is the viscous stress due to deformation of the entire mixture of slurry and the coarse particles, τ_c is the shear stress due to inelastic particle collision, τ_k the kinetic stress due to plunging of particles from a layer into other layers, τ_s the static Mohr-Coulomb shear stress due to enduring contact motion between the particles, τ_t the turbulent mixing stress, p_c the pressure due to particle collision, p_s the static pressure due to enduring frictional contact between particles, and p_h is the isotropic pressure in the interstitial fluid. If $(\tau_y + \tau_\mu), (\tau_c + \tau_s)$ and $(\tau_k + \tau_t)$ are written as τ_v, τ_g and τ_m, respectively, τ_v is considered to represent the shear stress shared by visco-plasticity of the material, τ_g is that shared by contact of the particles and τ_m is that shared by mixing of the material. Three extreme cases in which only one of these three shearing stresses is dominant can be considered; when τ_v dominates flow is the viscous type, when τ_g dominates flow is either the stony inertial type or the quasi-static shearing motion, and when τ_m dominates flow is the turbulent muddy type.

Many previous experimental and theoretical considerations on the constitutive relations of granular flows (e.g. Bagnold [1], Savage [8]) confirm the following forms:

$$\tau_c = \sigma d^2 f_1(C)\gamma^2 \tag{5.3}$$

$$\tau_k = \sigma d^2 f_2(C)\gamma^2 \tag{5.4}$$

$$\tau_\mu = \mu f_3(C)\gamma \tag{5.5}$$

where σ the particle density, γ the shear rate, μ the viscosity of the interstitial fluid that incorporates with the fine particles, and $f_1(C), f_2(C), f_3(C)$ are the functions of the coarse particle concentration C, respectively. In this paper the term coarse particle means the particles coarser than that constitutes the interstitial slurry. The shear stress due to macro-turbulence is given by

$$\tau_t = \rho_a l^2 \gamma^2 \tag{5.6}$$

where ρ_a is the apparent density of the material, and l is the mixing length. From these formulae, if τ_y is negligibly small in comparison to τ_μ as is justified in an actual debris flow, the ratio τ_μ/τ_c is given as

$$\frac{\tau_\mu}{\tau_c} = \frac{\mu}{\sigma d^2 \gamma} \frac{f_3(C)}{f_1(C)} \equiv \frac{1}{\mathbf{Ba}} \frac{f_3(C)}{f_1(C)} \left\{ \left(\frac{C_*}{C} \right)^{1/3} - 1 \right\}^{-1/2} \tag{5.7}$$

where $\mathbf{Ba}(= \sigma d^2 \gamma \{(C_*/C)^{1/3} - 1\}^{-1/2}/\mu)$ is called Bagnold number, and C_* is the coarse particle concentration when packed. The ratio (τ_μ/τ_c) becomes large (effect of particle collision becomes small in comparison to the effect of viscosity) when Bagnold number becomes small. Bagnold [1] showed by his experiment that the macro-viscous flow regime appears when Bagnold number is smaller than about 40. According to Takahashi and Tsujimoto [11] τ_k is far smaller than τ_c if C is larger than about 0.3, so that for the case of larger solids concentration the ratio (τ_m/τ_μ) can be approximated by (τ_t/τ_μ), and that ratio is given as

$$\frac{\tau_t}{\tau_\mu} = \frac{\rho_a l^2 \gamma}{\mu f_3(C)} \sim \frac{\rho_a U h}{\mu f_3(C)} = \mathbf{Re} \tag{5.8}$$

where $\mathbf{Re}$ is the Reynolds number, U the mean velocity of flow, h the depth of flow and $\mu f_3(C)$ is the apparent viscosity of the entire debris flow material, in which $f_3(C)$ increases with increasing concentration. Therefore, the effect of turbulence becomes minimal when Reynolds number becomes small.

Thus, viscous debris flow appears under the condition that the small Bagnold number, high solids concentration and small Reynolds number. This condition is satisfied when the concentration of the fine cohesive particles in the interstitial fluid is large (μ becomes large so as to make Bagnold and Reynolds numbers small), the coarse particle concentration is also large and surge. The discharge from the hopper gradually decreased with reduction of the material in the hopper, and shortly after all the material in the hopper was released a certain thickness of deposit remained on the bed of the flume nearly parallel to the bed surface. The same quantity of the material as the first surge was prepared in the hopper, and five to six minutes later the termination of the first surge, the second surge was introduced onto the deposit left after the first surge. In some cases, the third, fourth and fifth surges were produced with the same procedure. The surface level of flow was measured at the two positions; 3m and 6m from the upstream end, using the laser deflection sensors. The discharge variation supplied from the hopper was calibrated by capturing the discharge directly with a calibration box.

Experimental material

Well mixed characteristics of the material in Jiangjia Gully was imitated by mixing silica sand and kaolin except for the component larger than 10mm.

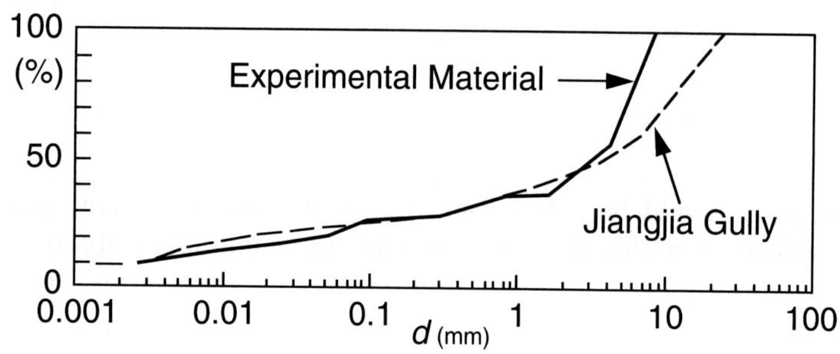

Fig. 5.6. Particle size distributions of the material in Jiangjia Gully and in the experiments.

The particle size distribution of the material in Jiangjia Gully (the component smaller than 20mm) and in the experimental material are compared in Figure 5.6. Although debris flows in Jiangjia Gully are said to flow with the solids concentration in water as much as 0.85 in volume (Wu et al. [12]), debris flows in the experiments could not reach the outlet of the flume if the solids concentration was higher than 0.58, and if it was lower than 0.55 the material easily segregated in the hopper before feeding into the flume. Therefore, many experimental runs were conducted under the nearly constant solids concentration in water from 0.56 to 0.57 ($\rho_a = 1.94\text{g/cm}^3$). The material in Jiangjia Gully lacks the fraction whose particle diameter is about 0.1mm, and the particles smaller than 0.1mm may be considered as the material that compose slurry as the interstitial fluid. Similarly, if the particles smaller than 0.1mm in the experimental material are considered to compose slurry, the slurry has the apparent density of 1.38gcm^{-3} (volume concentration of the solids is 0.226). Then, the median diameter of the coarse particle fraction is about 4.5mm. The volume concentration of the coarse particle fraction in slurry ia about 0.44. Slurry in the experimental material itself was well modeled as a Bingham fluid having the viscosity and the yield strength 0.036Pa·s and 11.8Pa, respectively.

These flume experiments by no means aimed to do a reduced scale model tests of the Jiangjia Gully debris flow, but the experiments were conducted under **Re**$\simeq$ 10 and **Ba**$\simeq$ 40.

5.4.2 General characteristics disclosed by the experiments

Apparent viscosity of the material

If the thickness of deposit left after passage of debris flow surges is assumed to correspond to the yield strength of the material, τ_y is given by $\rho_a gh \sin\theta$. The experimental results showed that thus obtained τ_y decreased with increase in channel slope. The yield strength is, however, a property of the material that must not change depending on the channel slope. This contradiction was derived from the assumption that the thickness of deposit corresponds to the yield strength. This is considered as one of the evidences to deny the applicability of the Bingham fluid model.

If the material is assumed as a Newtonian fluid, the apparent viscosities were obtained as shown in Figure 5.7. The experiments were conducted in narrow concentration range and at about the concentration of 0.54 the obtained results much scattered, because this was almost the least concentration to be able to get homogeneity within the entire material. Figure 5.7 shows the apparent viscosity of the entire material is about hundred times of the viscosity of slurry. Objection for regarding the material as a Newtonian fluid may arise, because Newtonian fluid will not stop as long as the channel within which the fluid flows has a certain slope. For this issue discussion will be developed later.

Behavior of flow

The first surge on the rigid bed leaves deposit of a certain thickness. The second surge travels on the already deposited layer, and it erodes and entrains the deposit in continuing its running down. As shown in Figure 5.8, if channel slope is steeper than 11°, the thicknesses of deposit after passage of the first and the second surges do not change much. But, if the channel slope is flatter than 9°, the thickness seems to increase surge by surge at least in the early stage of repetition of surges. In this case, the sediment composition in the flow discharging from the outlet of the flume becomes smaller than in the supplied material; i.e., the larger particles deposit more than the fine particles. In the case of 9° channel slope, the thickness of deposit increased surge by surge until fourth surge, but it ceased to increase in the fifth surge. Although existence of such a limit in deposit thickness is not yet clear in the experiments due to shortage of experimental runs, in Jiangjia Gully it surely deposit layer was faster than that on the rigid bed by about two times. Note that the discharge hydrographs supplied upstream for the first and second surges were almost identical. The translation velocity of the forefront on the deposit layer was faster by 1.5 to 2 times than the mean velocity of flow measured at a little behind the front part. This suggests that the front of the second surge has more or less the characteristics of wave.

Taking the first surge on 11° slope channel as an example, the velocity distribution, the flow depth and the deposit thickness in the respective time intervals are shown in Figure 5.10. The closed circles in the figure show the velocity of the coarse particles at their vertical positions that is obtained by the analysis of the high-speed video records. The deposit layer gradually thickens and the vertical velocity gradient becomes smaller with decrease in the flow depth. These characteristics cannot be explained by the Bingham fluid model, in which the deposit layer abruptly appears when flow depth decreased to the thickness of plug that necessarily exist in the upper part of flow. As is evident in the figure, there is no plug in flow. In the figure the situations up to the time when the thickness of deposition reaches 0.8cm are demonstrated. In this experimental run, however, the final thickness became 2cm by the prolonged very shallow and slow flow.

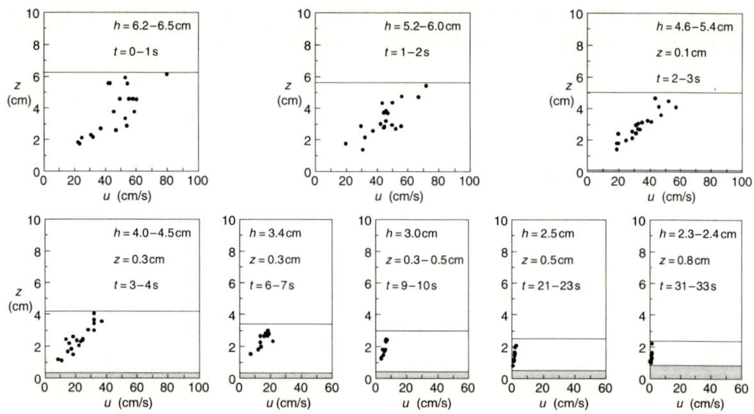

Fig. 5.10. Velocity distribution, flow depth and deposit thickness versus time.

5.4.3 Mechanism to disperse the coarse particles in a highly viscous newtonian fluid

Why the coarse particles disperse in viscous debris flow is an interesting theme from the mechanical point of view. Some discussions have been done concerning this problem. For example, Takahashi [10] discussed that the approach of the two particles respectively embedded in the two adjacent shearing planes generates flow in the interstitial fluid perpendicular to the shearing planes and the flow becomes a cause to disperse the particles. Iverson [4] claimed that, even all the coarse particles in a debris flow are irreversibly settling toward the bottom, the time necessary to settle down all the particles is normally much longer than the life span of the debris flow itself. Therefore, the particles are apparently dispersed as long as the debris flow continues. But, this hypothesis may fail in explaining the phenomenon that the once deposited layer from a surge is remobilized and entrained by another surge.

Resuspension of heavy particles in laminar flow has attracted interest of chemical engineers rather than geophysical researchers. Phillips et al. [6] among others developed a theory of particle dispersion in laminar flow. Here the particle dispersion in viscous debris flow is treated based on their considerations. Consider a situation that two particles embedded in adjacent shearing surfaces move past one another. The number of encounters experienced by a test particle will scale as γC. Therefore, the variation in the frequency of particle encounter over a distance $O(d)$ is given by $d\nabla(\gamma C)$. If the particle migration velocity that takes place perpendicular to the main flow direction is assumed linearly proportional to this variation in the frequency of particle encounter and if each of these two-body interactions give rise to a displacement of $O(d)$, the particle flux perpendicular to the main flow direction is given by

$$\boldsymbol{N}_c = -K_c d^2 (C^2 \nabla\gamma + C\gamma\nabla C) \tag{5.9}$$

where K_c is a constant.

Takahashi [10] inferred that, in a space the particles are homogeneously dispersed, the force operating on a particle perpendicular to the main flow direction would be proportional to $d^2 f'(C)\gamma$ where $f'(C)$ is a function of C. The variation in the force over a distance of $O(d)$ is then given by $d^2 f'(C)\nabla\gamma$. Extension of this consideration to a non homogeneous space gives that the force variation over a distance $O(d)$ is proportional to $d^2 f'(C)\nabla(\gamma C)$. This force variation should cause a net displacement of the particle. Because this force is induced by the drag force of the interstitial fluid, the following relationship should be satisfied:

$$d^2 f'(C)\nabla(\gamma C) \propto (1/2)\rho C_D(\pi/4)d^2 w^2 \tag{5.10}$$

where w is the net local velocity perpendicular to the main flow in the interstitial fluid that is induced by squeezing and relaxation of the interstitial fluid, and C_D is the drag coefficient. For a laminar flow, $C_D = 24/\mathbf{Re} = 24\mu/(\rho w d)$, then

$$w \propto \frac{1}{3\pi\mu} d f'(C)\nabla(\gamma C) \tag{5.11}$$

where μ is the viscosity of the interstitial fluid. Assuming $f'(C) \sim C$ and the particle velocity immediately assimilates with that of the ambient fluid, the

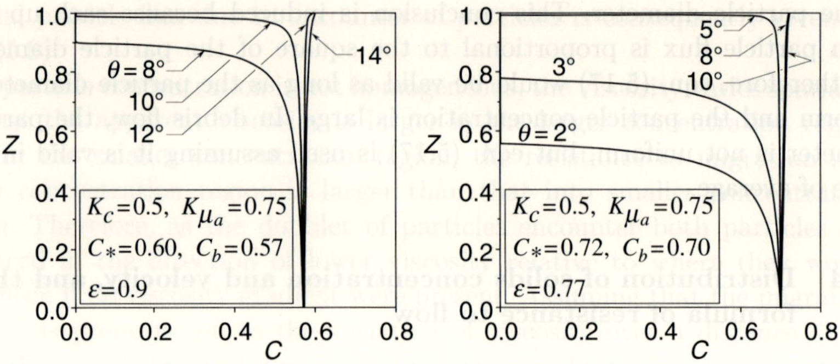

Fig. 5.11. Distributions of the coarse particle concentrations on various channel slopes.

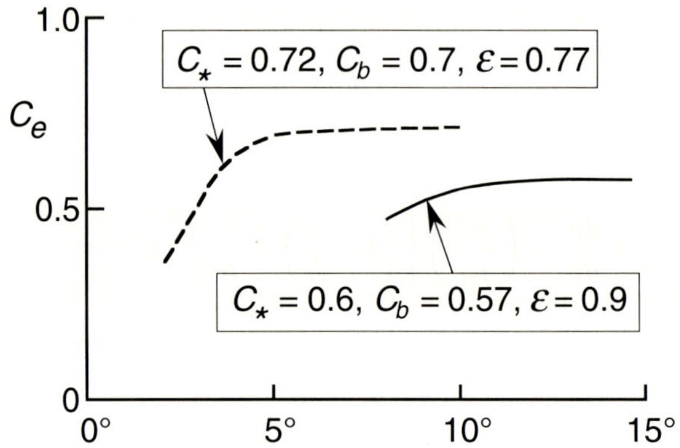

Fig. 5.12. Equilibrium coarse particle concentrations versus channel slopes.

The coarse particle concentration becomes large and uniform as ε becomes small. Namely, the denser the slurry, the flatter area the debris flow can reach. If a debris flow contains the particles with very wide range of sizes from very fine clay to large stones, it is difficult problem to decide up to what size particles can be considered as the material to form the interstitial slurry. The typical viscous debris flow that arises at Jiangjia Gully has very wide particle size distribution, and it passes through a mild slope reach with very high solids concentration and velocity. Therefore, it must have small ε value, perhaps even the particles larger than the clay size are effective to make the interstitial slurry. The equilibrium coarse particle concentration for a given channel slope is obtained by integrating eqn. (5.21). The obtained results corresponding to the conditions in the experiments and in Jiangjia Gully are shown in Figure 5.12.

The velocity distribution is given by the integration of eqn. (5.20), but if one simply assumes a uniform concentration throughout the depth and takes the boundary condition at $Z = 0, u' = 0$ into account, one obtains

$$u' = \frac{\rho\sqrt{gh}h \sin\theta}{\mu_a}(1 + \varepsilon\bar{C})\left(Z - \frac{1}{2}Z^2\right) \tag{5.22}$$

where $\bar{C}$ is the cross-sectional mean concentration of coarse particles and $u' = u/\sqrt{gh}$. The velocity distributions by eqn. (5.22) are compared with the experimental data in Figure 5.13. This figure shows that not only the velocity distribution obeys the Newtonian laminar flow equation, but also the absolute velocity values are well predicted by the theory herein introduced. In the calculation of eqn. (5.22) μ_a is obtained from eqn. (5.15), in which the viscosity of slurry, μ, is obtained by conducting the experiments to measure the viscosity of slurry, and for C and C_* values the entire solids concentration including the coarse and fine particles is used. The viscosity obtained from eqn. (5.15) is compared with the experimental results in Figure 5.7.

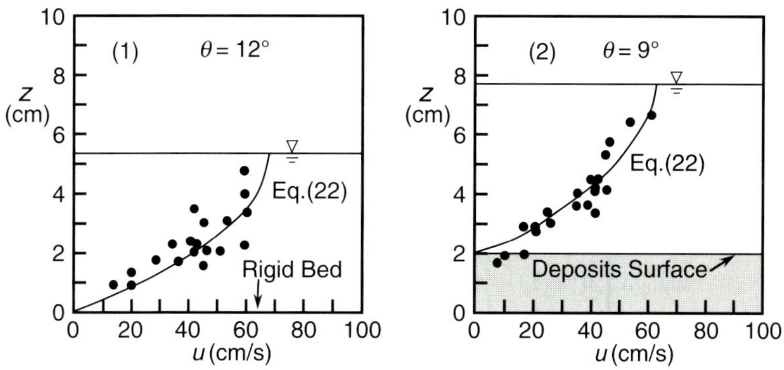

Fig. 5.13. The calculated velocity distributions (parabolic curves) and the experimental results (circles). (1) On the deposit layer, (2) on the rigid bed.

Integration of eqn. (5.22) gives the formula of the resistance to flow as following:

$$\frac{U}{u_*} = \frac{1}{3}\frac{\rho u_*}{\mu_a}(1 + \varepsilon\bar{C})h \tag{5.23}$$

where U is the cross-sectional mean velocity.

Let us apply the theory to the Jiangjia gully debris flow. According to the observational data (Wu et al. [12]), volume concentration of the fine particles in the slurry is about 30% and the volume concentration of the coarse particles in the slurry is about 55%. Thus, the densities of slurry and the entire debris flow material are, respectively, $\rho = 1.5\text{g cm}^{-3}$ and $\rho_a = 2.13\text{g cm}^{-3}$; i.e., $\varepsilon = 0.77$. The total solids concentration in the debris where **Re** is the Reynolds number; $(= \rho_a U H_f/\mu_a)$, and because the flow is laminar $f = 6/\mathbf{Re}$ is satisfied.

Let us assume the velocity of the hatched volume becomes zero when z_b coincides with h_0 and during that process the velocity decreases linearly. Then, the depositing speed $i(= dz_b/dt)$ is, writing the time necessary for this process as T, $i = (h_0 - z_b)/T$. Because $T = -iU_0/(h_0 - z_b)$, the following formula is obtained:

$$\frac{dU_0}{dt} = -\frac{U_0}{h_0 - z_b} i \tag{5.29}$$

The relationship between the mean velocity of the debris flow; U, and U_0 depends on the velocity distribution, but, here, simply the following relation is assumed:

$$U_0 = \alpha U \quad (\alpha < 1) \tag{5.30}$$

Consequently,

$$i = \frac{gH_f \sin\theta}{2\alpha U}\left\{\frac{(\sigma - \rho_a)(C - C_e)\tan\phi}{\rho_a \tan\theta} - 1 + \frac{3\mu_a U}{\rho_a g H_f^2 \sin\theta}\right\} \tag{5.31}$$

In laminar flow, the second and third terms in the curly bracket in eqn. (5.31) are equal. Therefore,

$$i = \frac{3\mu_a(\sigma - \rho_a)(C - C_e)\tan\phi}{2\alpha H_f \rho_a^2 \tan\theta} \tag{5.32}$$

The relationship between the channel slope and the equilibrium coarse particle concentration was given in Figure 5.12. The coarse particle concentration in the material for the laboratory flume experiment was about 0.425. Therefore, if the relationship given in Figure 5.12 is quantitatively correct, eqn. (5.32) necessarily leads to the conclusion that no deposition will occur in the channel steeper than 8°. This is not accord with the experiments in which deposition occurs even on a far steeper slope. This fact, in turn, suggests that more strict discussions on the coefficient K_c, K_{μ_a}, etc. are necessary. Anyway, to explain the depositing tendency in the experiments a little smaller C_e must be given.

According to the experiment, when a debris flow surge front passes over the layer deposited by a previous surge, the entire layer begins to move again together with the arriving surge. This phenomenon happens because the deposit still has almost the same properties as the moving material. The freshly deposited layer is barely stable due to a weak binding among the coarse particles, but, it is easily destabilized by impulse of the arriving surge. The viscous debris flow material is originally so highly concentrated that the volume change from moving stage to deposit is very small.Removal of the deposit by an arriving surge is confirmed in Jiangjia Gully. There, sometimes, even when no surge arrives, minimal disturbance leads to dilation of the deposit layer giving rise to a bore. The deposit maintains its original very soft condition over several days and it is difficult to walk on it. The deposit gradually loses water by evaporation and infiltration, and becomes stiff.

5.5 One-Dimensional Routing of the Debris Flow Surges

One-dimensional routing of the debris flow surges is fulfilled by solving the following equations of continuity and momentum conservation.

$$\frac{\partial H_f}{\partial t} + \frac{\partial (UH_f)}{\partial x} = -i \tag{5.33}$$

$$\frac{\partial (UH_f)}{\partial t} + 1.2\frac{\partial (UH_f)}{\partial x} = gH_f \sin\theta_b - gH_f \cos\theta_b \frac{\partial H_f}{\partial x} - \frac{9\mu_a^2 U^2}{\rho_a^2 g H_f^3 \{\sin\theta_b - \cos\theta_b(\partial H_f/\partial x)\}} \tag{5.34}$$

where θ_b is the original bed slope. The deposit layer is assumed to be immediately eroded to the depth of original bed as soon as a surge front arrives. Then deposition proceeds at the speed i given by eqn. (5.32).

The experimental data for the case of 11° in the channel slope are compared with the calculation using Eqs. (5.33) and (5.34) in Figure 5.16. In the calculation following values are used: $C_e = 0.41$, $\alpha = 0.9$, $\tan\phi = 0.7$, $\rho_a = 1.92\text{g cm}^{-3}$, $\sigma = 2.65\text{g cm}^{-3}$, $\rho = 1.38\text{g cm}^{-3}$ and $\mu_a = 6.2\text{Pa·s}$. As the upstream boundary condition the discharge hydrograph of a surge depicted in Figure 5.17 is given.

In the experiment, the temporal variations of the surface stage at the upstream station (3m downstream from the debris flow supply point) and at the downstream station (3m downstream from the upstream station) are recorded. Figure 5.16(a) shows the results for the first surge, which traveled on a rigid bed (no deposit layer exists before arrival of the surge). The translation velocities of the front both in the calculation and in the experiment are the same 73.7cm s^{-1} and it takes about 4 seconds to travel between the two stations. Apart from the calculated front stage being a little higher than the experimental result, the general tendency of stage variation is well reproduced by the calculation. With time, both calculated and experimented stages approach constant heights at respective stations. This means the flow has already stopped. The calculated time of stopping and the thickness of deposit coincide well with the experiment.

Figure 5.16(b) shows propagation of the second surge that traveled on the deposit made by the first surge. There are some differences in the front arrival time between the calculation and the experiment, but, it is probably due to difference in setting of the starting time, because the time lags between the two stations are almost the same; about 2.7 seconds. Stage variation in the experiment has dual peaks, whereas in the calculation it has only one peak. This probably has happened due to difficulty in controlling supply discharge in the experiment. The hydrographs of the first and second surges in the calculation are identical. Except for such discrepancy, the calculation

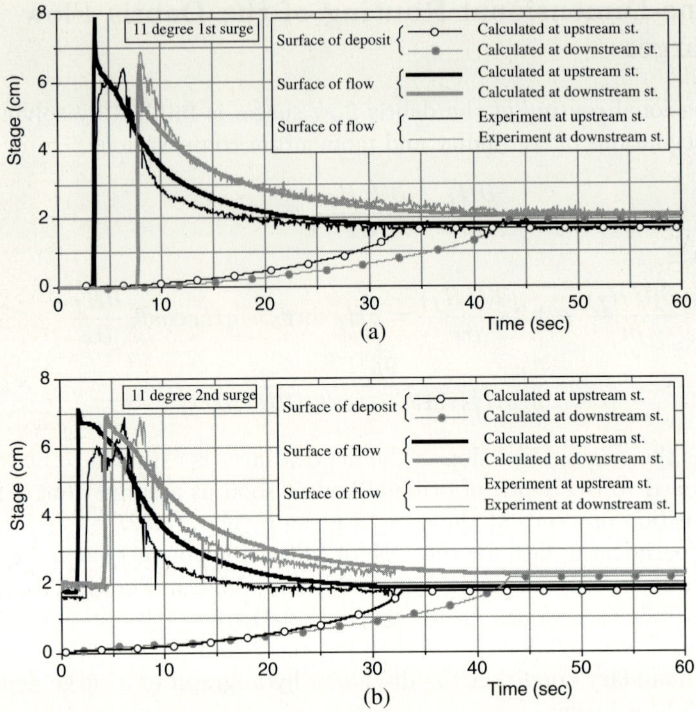

Fig. 5.16. Comparison between the calculated and the experimented surface stages of flow and deposit for the case of 11° channel slope. (a): the first surge on the rigid bed, (b): the second surge on the deposit made by the first surge.

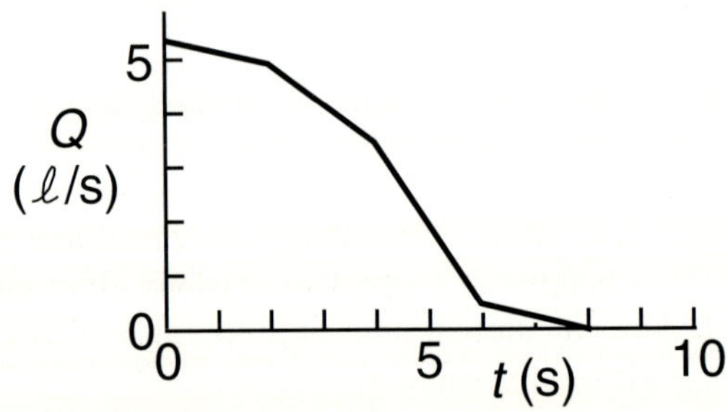

Fig. 5.17. The hydrograph of debris flow supplied from the hooper.

generally explains well the experimental results. The time required for the first surge to travel the 3m reach is 4 seconds and that for the second surge is only 2.7 seconds. The translation velocity of the first surge is almost equal to the mean cross-sectional velocity at the peak. Therefore, existence of the deposit contributes, in this case, to making the front velocity faster than the mean velocity of the following part.

Note that the final stages of the surface of the deposit after the first surge and after the second surge are almost the same. This is one of the peculiar behaviors in Jiangjia Gully as well.

5.6 Conclusion

In this paper, at first, debris flows were classified in the mechanical point of view, and the condition under which viscous debris flows appear was given. Namely, a viscous debris flow appears when the concentrations of the fine cohesive particles in the interstitial fluid and of the coarse particles in the entire material are simultaneously large and, in addition, the diameter of the coarse particles is not very large.

Next, referring to the typical viscous debris flows which occurred in the Jiangjia Gully basin, Yunnan, China, the characteristics of viscous debris flow were made clear, and the main issues in the mechanics of viscous debris flow were given. The intermittency which is the most remarkable behavior of Jiangjia Gully debris flow was attributed to the occurrence of the numerous landslides widely distributed in the basin. A big scale landslide which occurred in the Harihara River basin, Kyushu, Japan revealed that the upper part of the slid soil block moved as a block, but the lower part liquefied and transformed into debris flow, and it flowed over the soil block. A conceptual process model for occurrence of viscous debris flow by the development of a liquefied layer in the neighborhood of the boundary between the moving soil block and the rigid ground surface was proposed.

A Newtonian fluid model introduced here can quantitatively explain the distinguishing characteristics of viscous debris flow such as the high competence for transporting dense solids concentration in gentle slope channel, the nearly uniform distribution of solids throughout the depth, the fast translation velocity of the surge front, and the deposition and the erosion on the arrival of a surge. Different from the Bingham fluid model, it explains the appearance of standstill layer after passage of a surge even no plug exists in the flow.

5.7 Acknowledgement

This work has been done with the help of Nakagawa, H., Satofuka, Y., and several students. The observational study at Jiangjia Gully had been done for about eight years as a joint research project between the Disaster Prevention Research Institute, Kyoto University and the Chendu Institute for Mountain Disaster and Environment, Chinese Academy of Sciences. Many researchers who were concerned with this project and the financial support of the Ministry of Education, Science, Sports and Culture, Japan are highly appreciated.

References

1. Bagnold, R.A. (1954) Experiments on a gravity-free dispersion of large solid spheres in a Newtonian fluid under shear. Proc. Roy. Soc. London A, **225**, 49-63.
2. Fleming, R.W., Ellen, S.D., Algus, M.A. (1989) Transformation of dilative and contractive landslide debris into debris flows. An example from Marin County, California. Engineering Geology, **27**, 201-223.
3. Huang, X., Garcia, M.H. (1998) A Herschel-Bulkley model for mud flow down a slope. J. Fluid Mech., **374**, 305-333.
4. Iverson, R.M. (1997) The physics of debris flows. Reviews of Geophysics, **35** 3, 245-296.
5. Krieger, J.M. (1972) Rheology of monodisperse lattices. Adv. Coll. Int. Sci., **3**, 111-136.
6. Phillips, R.J., Armstrong, R.C., Brown, R.A., Graham, A.L., Abbot, J.R. (1992) A constitutive equation for concentrated suspensions that accounts for shear-induced particle migration. Physics of Fluids A, **4**(1), 30-40.
7. Sabo Technical Center (1997) Actual situation of the sediment disaster in 1997 (in Japanese). Sabo Tech. Center.
8. Savage, S.B. (1984) The mechanics of rapid granular flows. Adv. Appl. Mech., **24**, 289-376.
9. Takahashi, T. (1991) Debris flow. Balkema, Rotterdam.
10. Takahashi, T. (1993) Fluid mechanical modelling of the viscous debris flow. Proc. Pierre Beghin International Workshop on Rapid Gravitational Mass Movements, 313-321.
11. Takahashi, T., Tsujimoto, H. (1997) Mechanics of grannular flow in an inclined chute (in Japanese with English abstract). J. Hyd. Coast. Environ. Eng., JSCE, 565/II, **39**, 57-71.
12. Wu, J., Kan, C., Tian, L., Zhang, S. (1990) Observational study of the debris flows in the Jiangjia gully, Yunnan (in Chinese). Science Publishing Co., China.

6 Transverse Slope of Bed and Turbid-Clear Water Interface of Channelized Turbidity Currents Flowing around Bends

Gary Parker[1], Jasmin Imran[2], Carlos Pirmez[3]

[1] St. Anthony Falls Laboratory University of Minnesota, USA
[2] Department of Civil and Environmental Engineering University of South Carolina, USA
[3] Exxon Production Research Company Houston, Texas USA

6.1 Abstract

Turbidity currents are sediment-laden bottom flows in lakes and the ocean that derive their momentum from the force of gravity acting on the sediment held in suspension. These currents are capable of creating highly meandering channels extending over hundreds or thousands of kilometers. The planforms of these channels show a remarkable resemblance to those of meandering rivers. In order to understand the flow of turbidity currents in sinuous channels a simple analysis of steady, streamwise-uniform flow around a bend of low, constant curvature is considered. The analysis is applied to both a river and a channelized turbidity current. In both cases the bed of the channel is sand, and both sand and mud are carried in suspension. The sand requires turbulence in order to be held in suspension, but the mud is sufficiently fine to allow it to be carried as wash load. The flow is assumed to be Froude-subcritical, and in the case of a turbidity current a relatively sharp interface between turbid water and clear water above is assumed. The analysis focuses on the processes that maintain a) the transverse water surface or interfacial slope associated with superelevation around bends and b) the transverse bed slope created by scour on the outside and fill on the inside of bends. In the case of rivers, the processes maintaining these two transverse slopes can be decoupled, allowing for the use of standard formulations. In a turbidity current, however, the processes may be strongly coupled, especially when the concentration of sand in suspension decreases strongly in the vertical.

6.2 Introduction

The most common river planform morphology is that of the meandering channel. A particularly striking image of a meandering river is that of Figure 6.1. Here such meanders are referred to as "subaerial", in that they are created by the interaction water flow and sediment transport in rivers under air (i.e. the atmosphere).

Fig. 6.1. Meandering river in Siberia. Photo courtesy A. Zaitsev.

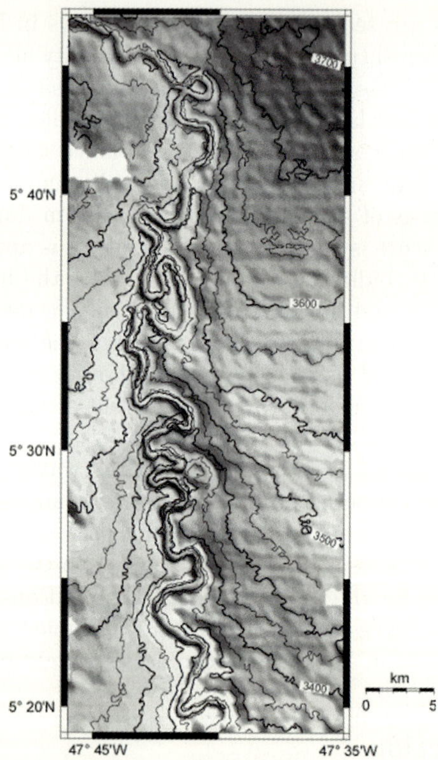

Fig. 6.2. Meandering channel on the Amazon submarine fan. From Pirmez [12].

The present paper is, however, concerned with another type of meandering channel, one with a striking resemblance to river meanders but in a radically different environment. The channels in question can be found at the bottom of the ocean and some lakes. They are formed by the action of turbidity currents, i.e. sediment-laden bottom currents that derive their driving force from the excess weight due to the presence of sediment. For this reason these channels can be said to display "subaqueous" meandering, in that the flows that create them consist of turbid water immersed in clear water. When these channels are found in the ocean, they may be said to describe "submarine" meandering. An image of submarine meandering that is every bit as striking as subaerial meandering is shown in Figure 6.2. Here the term "submarine meandering" is used so as to include deep-water meandering channels in lakes as well.

Submarine meandering channels possess both remarkable similarities to and notable differences from subaerial meandering channels. Here the analysis is focused on a relatively simple configuration: steady flow around a bend of constant curvature that is also uniform in the streamwise direction. Results are presented in parallel for both rivers and channelized turbidity currents.

The channel configuration to be studied is shown schematically in Figure 6.3, in which s denotes the streamwise direction, n denotes the transverse direction (directed such that n is positive on the outside of the bend) and z denotes a coordinate upward normal (nearly vertical) from the bed. The bend has constant centerline radius of curvature r_o and channel half-width b.

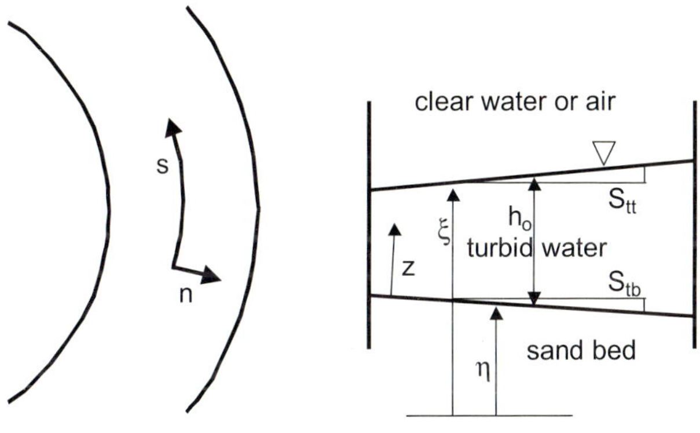

Fig. 6.3. Planform (left) and cross-section (right) definition diagrams.

In the case of rivers it is well known that the flow superelevates around bends. That is, the centrifugal force generated by the bend is balanced by an air-water interface, or water surface that is high on the outside and low on the inside. The magnitude of the resulting transverse water surface slope is denoted on Figure 6.3 as S_{tt}. In addition, the secondary flow associated with a bend causes bed scour on the outside of the bend and deposition on the

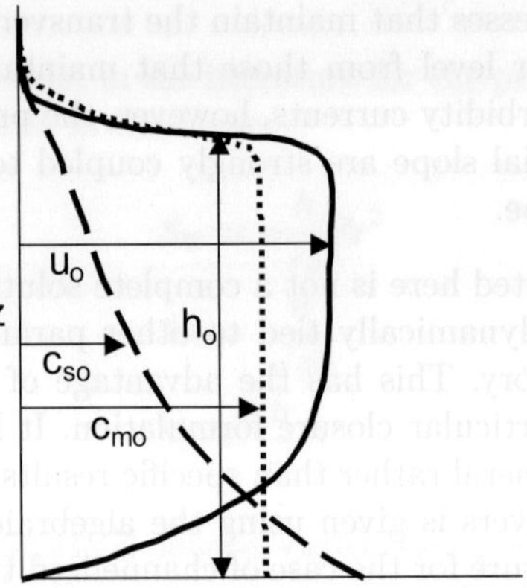

Fig. 6.5. Definition diagram for the vertical structure of a turbidity curren t.

6.3 Setup

The case under consideration is that of a curved c hannel with constant half-width b and radius of curvature in the streamwise direction, as illustrated in Figure 6.3. The bottom of the channel has constant transverse slope S_{tb}. The sidewalls of the channel are taken to be vertical and infinitely high. Channel curv ature is assumed to be small compared to half-width; that is,

$$\frac{b}{r_o} \ll 1 \tag{6.2a}$$

The above assumption allows a treatment that is linearized for small curvature. The basis for the analysis presented here is the linear formulation of P ark er and Johannesson [10] for riers.

In both the subaerial and submarine cases the flow is approximated to be steady and uniform in the streamwise and normal directions. Let h_o denote the centerline depth of flow in the channel in the subaerial case, and the centerline thickness of flow from bed to interface in the submarine case. It is further assumed that the flow is slender in the sense that

$$\frac{h_o}{b} \ll 1 \tag{6.2b}$$

The above approximation allows the neglect of sidewall effects everywhere in the flow except in the immediate vicinity of the sidewalls themselves.

As flo w traverses the bend, cen trifugal effects drive a superelevation of flow on the outside of the bend. Here this superelevation is simply described in terms of a top transverse slope S_{tt}.

The suspension is tak en to be dilute. Let ρ denote the density of clear w ater, ρ_s denote the material density of sediment (sand or mud) and c_s and

c_m denote the volume concentration of sand and mud, respectively, averaged over turbulence. It is thus assumed that

$$c_s + c_m \ll 1 \tag{6.3}$$

Let R denote the submerged specific gravity of the sediment, here defined as

$$R = \frac{\rho_s}{\rho} - 1 \tag{6.4}$$

For natural sediment R is typically close to 1.65. In a river the downslope driving force per unit weight of flow mixture is given as

$$gS_s[1 + R(c_s + c_m)] \cong gS_s \tag{6.5}$$

where g denotes the acceleration of gravity and S_s denotes the downstream bed slope. The approximation is valid for a dilute suspension. The turbidity current, however, flows only because of the presence of suspended sediment, which renders it heavier than the surrounding clear ambient water. As a result the downslope driving force per unit weight is given by the term

$$RgS_s(c_s + c_m) \tag{6.6}$$

It is seen via a comparison of (6.5) and (6.6) that for the same bed slope and the same dilute concentrations of sand and mud the driving force of a turbidity current is vastly reduced compared to that of a river.

6.4 Primary Flow

Here the limiting case as $r_o \to \infty$, i.e. flow in a straight channel is considered. In both the subaerial and subaqueous cases the flow velocity averaged over turbulence is denoted as $u(z)$. The vertical mixing of both streamwise momentum and suspended sediment is represented in terms of a kinematic eddy viscosity $\epsilon(z)$. Here $\epsilon(z)$ is taken to be an arbitrary function, which nevertheless must take substantially different forms in the subaerial and submarine cases. The specification of a form for ϵ requires a specific turbulence closure theory. The assumption that the eddy diffusivity of momentum and suspended sediment should be identical corresponds to a Schmidt number of unity.

6.4.1 River

The governing equation for streamwise, or primary momentum balance in a straight open channel is

$$0 = gS_s + \frac{d}{dz}\epsilon(z)\frac{du_o}{dz} \tag{6.7a}$$

In addition the bottom boundary condition (6.7b) takes the specific form

$$u_s = u_o|_{z=0} = \chi h_o \left.\frac{du_o}{dz}\right|_{z=0} \tag{6.19a}$$

where

$$\chi = \chi_1 - \frac{1}{3} \tag{6.19b}$$

$$\chi_1 = \frac{C_f^{-1/2}}{13} \tag{6.19c}$$

$$C_f = \left(\frac{u_{*o}}{\overline{u}_o}\right)^2 \tag{6.19d}$$

Here χ and χ_1 are order-one parameters, the former varying between 0.44 and 1.97 and the latter varying between 0.77 and 2.31 as $C_f^{-1/2}$ varies between 10 and 30. The above assumptions specify the following forms for f_{ru} and f_{rs}:

$$f_{ru} = \frac{\chi + \zeta - \frac{1}{2}\zeta^2}{\chi_1} \qquad f_{rs} = \frac{v_s h_o}{\epsilon} \frac{\exp\left(-\frac{v_s h_o}{\epsilon}\zeta\right)}{1 - \exp\left(-\frac{v_s h_o}{\epsilon}\right)} \tag{6.20a, b}$$

In addition, the relation between depth-averaged sand concentration $\overline{c}_{so}$ and the dimensionless sand entrainment parameter E is found from (6.13b) to be

$$\overline{c}_{so} = \frac{\epsilon}{v_s h_o}\left[1 - \exp\left(-\frac{v_s h_o}{\epsilon}\right)\right] E \tag{6.21}$$

García and Parker [2] provide examples of functional forms for E as a function of boundary shear stress and other parameters.

6.4.2 Channelized Turbidity Current

As noted above, the turbidity current is assumed to be subcritical in the bulk Richardson sense, and to have sufficiently strong stratification in the vicinity of the interface between turbid and clear water so as to allow for a neglect of interfacial mixing at first order. The flow thickness $z = h_o$ of the turbidity current is defined to be the point above the velocity maximum at which the magnitude of the shear stress τ_{io} takes its maximum value, as shown in Figure 6.5. Here τ_{io} is referred to as the interfacial shear stress, by definition given by the relation

$$-\epsilon \left.\frac{du_o}{dz}\right|_{z=h_o} = \frac{1}{\rho}\tau_{io} \tag{6.22}$$

A condition for the interface to be relatively sharp is a sufficiently high gradient Richardson number **Ri** there, i.e.

$$\mathbf{Ri} = -\left.\frac{Rg\frac{d(c_{mo}+c_{so})}{dz}}{\left(\frac{du_o}{dz}\right)^2}\right|_{z=h_o} \tag{6.23}$$

Here a formulation for turbidity currents that is valid over the range $0 \le z \le h_o$ is sought. The turbidity current above $z = h_o$ is neglected in accordance with the assumption of a sharp interface illustrated in Figure 6.5.

The formulation of momentum balance of (6.7a-c) for rivers takes the following form for turbidity currents:

$$0 = RgS_s(c_{mo} + c_{so}) + \frac{d}{dz}\epsilon(z)\frac{du_o}{dz} \tag{6.24a}$$

and

$$u_o(0) = u_s \tag{6.24b}$$

$$-\epsilon \left.\frac{du_o}{dz}\right|_{z=h_o} = \tau_{io} \tag{6.24c}$$

The formulations for mass balance of sand and mud remain unaltered from their forms for rivers. A reduction of the relations involved leads to the following solutions: u_o is given by the relation below,

$$u = h_o \int_0^{\zeta} \frac{1}{\epsilon}\left[RgS_s h_o \int_{\zeta'}^{1} (\bar{c}_{mo} + \bar{c}_{so}) d\zeta'' - \frac{\tau_{io}}{\rho}\right] d\zeta' \tag{6.25}$$

and c_{mo} and c_{so} are again given by the relations (6.10b) and (6.10c) respectively. In addition,

$$\frac{u_o}{\overline{u}_o} = f_{tu}(\zeta) \qquad \frac{c_{so}}{\bar{c}_{so}} = f_{ts}(\zeta) \qquad \frac{c_{mo}}{\bar{c}_{mo}} = f_{tm}(\zeta) \tag{6.26a – c}$$

where $\overline{u}_o, \bar{c}_{mo}$ and $\bar{c}_{so}$ are again given by (6.13), and

$$f_{tu} = \frac{\int_0^{\zeta} \frac{1}{\epsilon}\left[RgS_s h_o \int_{\zeta'}^{1} (\bar{c}_{mo} + \bar{c}_{so}) d\zeta'' - \frac{\tau_{io}}{\rho}\right] d\zeta'}{\int_0^1 \int_0^{\zeta} \frac{1}{\epsilon}\left[RgS_s h_o \int_{\zeta'}^{1} (\bar{c}_{mo} + \bar{c}_{so}) d\zeta'' - \frac{\tau_{io}}{\rho}\right] d\zeta' d\zeta} \tag{6.27a}$$

$$f_{ts} = \frac{\exp\left(-v_s h_o \int_0^{\zeta} \frac{d\zeta'}{\epsilon}\right)}{\int_0^1 \exp\left(-v_s h_o \int_0^{\zeta} \frac{d\zeta'}{\epsilon}\right) d\zeta} \tag{6.27b}$$

$$f_{tm} = 1 \tag{6.27c}$$

Note from the form of (6.24a) that in the case of a turbidity current the velocity profile is strongly coupled to the concentration profile. Although not explicit in the present formulation, even algebraic closures for ϵ link the concentration profile to the velocity profile, as was seen for the Engelund [1] closure applied above to rivers.

The above formulation allows the definition of a densimetric Froude number $\mathbf{Fr}_d$ and a bulk Richardson number $\mathbf{Ri}_b$ for the primary flow:

$$\mathbf{Fr}_d = \frac{\overline{u}_o}{\sqrt{Rg(\bar{c}_{mo} + \bar{c}_{so})h_o}} \qquad \mathbf{Ri}_b = \frac{Rg(\bar{c}_{mo} + \bar{c}_{so})h_o}{\overline{u}_o^2} \tag{6.28a, b}$$

The condition for a subcritical turbidity current is $\mathbf{Fr}_d < 1$ or equivalently $\mathbf{Ri}_b > 1$.

6.5 Secondary Flow

Here the case of secondary flow subject to the conditions (6.2a,b) is considered. The transverse bed slope S_{tb} is assumed to be positive when bed elevation η decreases in the positive n direction. The transverse water surface or interfacial slope S_{tt} is assumed to be positive when the elevation of the water surface or interface ξ increases in the positive n direction. Noting that

$$\xi \cong \eta + h \tag{6.29a}$$

it follows that

$$S_{tb} = -\frac{\partial \eta}{\partial n} \tag{6.29b}$$

$$S_{tt} = \frac{\partial \xi}{\partial n} = -S_{tb} + \frac{dh}{dn} \tag{6.29c}$$

These sign conventions are illustrated in Figure 6.3. The secondary flow itself is denoted as $v(z)$, and is evaluated everywhere except in thin layers near the sidewalls.

6.5.1 River

The equation of conservation of transverse momentum balance can be approximated by the following linearized form in the case of uniform flow in a bend satisfying (6.2a):

$$-\frac{u_o^2}{r_o} = -\frac{1}{\rho}\frac{dp}{dn} + gS_{tb} + \frac{d}{dz}\left(\epsilon \frac{dv}{dz}\right) \tag{6.30}$$

Here the pressure distribution can be approximated as hydrostatic, such that

$$p = \rho g(h - z) \tag{6.31a}$$

or thus

$$-\frac{dp}{dn} = -\rho g \frac{dh}{dn} = -\rho g(S_{tb} + S_{tt}) \tag{6.31b}$$

Substituting (6.31b) into (6.30), it is found that

$$-\frac{u_o^2}{r_o} = -gS_{tt} + \frac{d}{dz}\left(\epsilon \frac{dv}{dz}\right) \tag{6.32a}$$

The boundary conditions on (6.32a) can be written as

$$v(0) = v_s \tag{6.32b}$$

$$\epsilon \left.\frac{dv}{dz}\right|_{z=h_0} \tag{6.32c}$$

where v_s is a slip velocity analogous to the slip velocity u_s of the primary flow. In addition, conservation of flow mass requires the integral constraint

$$\int_0^{h_o} v dz = 0 \tag{6.32d}$$

The transverse boundary shear stress at the bed is given by the relation

$$\epsilon \left. \frac{dv}{dz} \right|_{z=0} = \frac{\tau_{nb}}{\rho} \tag{6.32e}$$

The problem can be reformulated in dimensionless terms as follows;

$$u_o^2 = \overline{u}_o^2 f_{ru}^2(\zeta) \quad \epsilon = \overline{u}_o h_o \hat{\epsilon} \quad \delta = \frac{h_o}{r_o} \quad S_{tt} = \delta \hat{S}_{tt} \quad v = \delta \overline{u}_o \hat{v} \qquad (6.33\text{a} - \text{e})$$

according to which (6.32a-d) can be rewritten as

$$-f_{ru}^2(\zeta) = -\mathbf{Fr}^{-2} \hat{S}_{tt} + \frac{d}{d\zeta}\left(\hat{\epsilon} \frac{d\hat{v}}{d\zeta}\right) \tag{6.34a}$$

$$\hat{v}(0) = \hat{v}_s \tag{6.34b}$$

$$\hat{\epsilon} \left. \frac{d\hat{v}}{d\zeta} \right|_{\zeta=1} = 0 \tag{6.34c}$$

$$\int_0^1 \hat{v} d\zeta = 0 \tag{6.34d}$$

Equations (6.34a-d) represent a single second-order relation subject to three boundary conditions. As a result it is possible to integrate to determine solutions to both $\hat{v}$ and the constant $\hat{S}_{tt}$. The results are found to be

$$\hat{v}(\zeta) = \hat{v}_s + \int_0^{\zeta} \frac{1}{\hat{\epsilon}(\zeta')} \left[\int_{\zeta'}^1 f_{ru}^2(\zeta'') d\zeta'' - \alpha_{rtt}(1 - \zeta') \right] d\zeta' \tag{6.35a}$$

$$\hat{S}_{tt} = \alpha_{rtt} \mathbf{Fr}^2 \tag{6.35b}$$

where

$$\alpha_{rtt} = \frac{\hat{v}_s + \int_0^1 \int_0^{\zeta} \frac{1}{\hat{\epsilon}(\zeta')} \int_{\zeta'}^1 f_{ru}^2(\zeta'') d\zeta'' d\zeta' d\zeta}{\int_0^1 \int_0^{\zeta} \frac{1}{\hat{\epsilon}(\zeta')} (1 - \zeta') d\zeta' d\zeta} \tag{6.35c}$$

As an example, the above scheme is implemented using the Engelund [1] closure procedure described above. In that approach the slip condition (6.34b) takes the specific form

$$\hat{v}_{ss} = \hat{v}|_{\zeta=0} = \chi \left. \frac{d\hat{v}}{d\zeta} \right|_{\zeta=0} \tag{6.36}$$

and

$$\gamma_{tt} = \int_0^1 \int_0^\zeta \frac{1}{\hat{\epsilon}(\zeta')} \int_{\zeta'}^1 X_t(\zeta'')d\zeta''d\zeta'd\zeta \tag{6.47c}$$

$$\gamma_{tb} = \int_0^1 \int_0^\zeta \frac{1}{\hat{\epsilon}(\zeta')} \int_{\zeta'}^1 X_b(\zeta'')d\zeta''d\zeta'd\zeta \tag{6.47d}$$

$$\gamma = \int_0^1 \int_0^\zeta \frac{1}{\hat{\epsilon}(\zeta')} \left[-\hat{\tau}_{ni} + \int_{\zeta'}^1 f_{tu}^2(\zeta'')d\zeta'' \right] d\zeta'd\zeta + \hat{v}_s \tag{6.47e}$$

Equation (6.47b) contains one of the two important results of the present analysis. In a subcritical turbidity current capable of forming a meandering channel, mud is likely to be distributed relatively uniformly in the vertical, whereas sand, which is much heavier, is likely to be characterized by a fairly strong concentration gradient in the vertical. It is this upward gradient of sand concentration that renders turbidity currents fundamentally different from rivers. If for example the current is assumed to contain no sand, so that f_{ts} can be dropped in (6.46b,c), and f_{tm} is evaluated using (6.27c) it is found that

$$X_b(\zeta) = \gamma_{tb} = 0 \tag{6.48}$$

so that the (6.47) reduces to

$$\hat{v}(\zeta) = \hat{v}_s + \int_0^\zeta \frac{1}{\hat{\epsilon}(\zeta')} \left[\int_{\zeta'}^1 f_{tu}^2(\zeta'')d\zeta'' - \alpha_{ttt}(1-\zeta') \right] d\zeta' \tag{6.49a}$$

$$\hat{S}_{tt} = \alpha_{ttt} \mathbf{Fr}_d^2 \tag{6.49b}$$

where

$$\alpha_{ttt} = \frac{\gamma}{\gamma_{tt}} = \frac{\hat{v}_s + \int_0^1 \int_0^\zeta \frac{1}{\hat{\epsilon}(\zeta')} \left[-\hat{\tau}_{ni} + \int_{\zeta'}^1 f_{tu}^2(\zeta'')d\zeta'' \right] d\zeta'd\zeta}{\int_0^1 \int_0^\zeta \frac{1}{\hat{\epsilon}(\zeta')}(1-\zeta')d\zeta'd\zeta} \tag{6.49c}$$

That is, the structure of the "solution" is identical to that for rivers save the generalization to densimetric Froude number and the added transverse interfacial stress term. Equation (6.49b) for superelevation reduces to the dimensioned form

$$S_{tt} = \alpha_{ttt} \frac{h_o}{r_o} \mathbf{Fr}_d^2 \tag{6.50}$$

While the present analysis does not allow for the evaluation of the constant α_{ttt}, it is likely order-one and sufficiently close to unity to allow for a computation of bend superelevation of muddy subcritical turbidity currents from (6.50) with α_{ttt} approximated to unity.

The case in which sand is a significant component of the turbidity current, however, offers a very different structure. In particular, it is seen from (6.47b) that the transverse surface slope is coupled to the transverse bed slope. As a result it is not possible to compute superelevation around bends in the absence of a predictive formulation for transverse bed slope.

6.6 Transverse Sediment Transport and Bed Slope

As noted above, it is assumed here that the mud in either a river or a turbidity current is not contained in, eroded from or deposited on the sand bed. For uniform bend flow the Exner equation of conservation of bed sand can be written in the form

$$(1-\lambda_p)\frac{\partial \eta}{\partial t} = v_s[c_o(0) - E] - \frac{\partial q_{nb}}{\partial n} \tag{6.51a}$$

Here the first term on the right-hand side of (6.51) denotes the net rate of settling of sand on the bed from suspension and the second term denotes the divergence of the transverse volume bedload transport rate per unit width q_{nb}. For steady flow (6.51a) reduces to

$$0 = v_s[c_{so}(0) - E] - \frac{dq_{nb}}{dn} \tag{6.51b}$$

The equation of conservation of suspended sediment in the water column for steady flow that is uniform in the streamwise direction is

$$\frac{dq_{ns}}{dn} = v_s[E - c_{so}(0)] \tag{6.52a}$$

where q_{ns} denotes the transverse volume transport rate of suspended sand per unit width, given by

$$q_{ns} = \int_0^{h_o} v c_{so} dz \tag{6.52b}$$

Between (6.51) and (6.52),

$$\frac{d}{dn}(q_{nb} + q_{ns}) = 0 \tag{6.53}$$

Integrating (6.53) under the condition of vanishing transverse bedload and suspended sand load at the sidewalls, it is found that

$$q_{nb} + q_{ns} = 0 \tag{6.54}$$

The above equation is a necessary condition for steady, uniform bend flow over a sand bed of a single grain size.

In considering transverse bedload transport in a bend, it is important to realize that the sand will have an inherent tendency to roll down any transverse slope. A variety of relations have been developed to predict this tendency (e.g. Parker and Johannesson [10]). In line with the linearized analysis presented above, the formulation of Parker and Andrews [9] is used here. When applied to steady, uniform bend flow it takes the form

$$\frac{q_{nb}}{q_{sbo}} = \frac{\tau_{nb}}{\tau_{sbo}} + \beta S_{tb} \tag{6.55a}$$

where

$$\beta = \beta^* \sqrt{\frac{\tau_c^*}{\tau_{so}^*}} \qquad \tau_{so}^* = \frac{\tau_{sbo}}{\rho R g D_s} \tag{6.55b}$$

In the above relation β^* is an order-one constant and τ_c^* denotes a critical Shields stress for the onset of motion. In general the solution to secondary flow in bends at linear level yields a solution for τ_{nb} of the form

$$\frac{\tau_{nb}}{\tau_{sbo}} = -\delta a_b \tag{6.56a}$$

where a_b is an order-one parameter. For example, in Parker and Johannesson's [10] implementation of the Engelund [1] closure it is found that

$$a_b = 7.51 \frac{\chi + \frac{2}{7}}{\chi_1} \tag{6.56b}$$

The transverse sand transport rate q_{ns} can be similarly related to the streamwise sand transport rate q_{sso} as follows:

$$q_{ns} = \frac{q_{ns}}{q_{sso}} q_{sso} = -\delta a_s q_{sso} \tag{6.57}$$

where according to (6.16b), (6.33e) and (6.52b)

$$a_s = -\frac{\int_0^1 \hat{v}(\zeta) f_{rs}(\zeta) d\zeta}{\int_0^1 f_{ru}(\zeta) f_{rs}(\zeta) d\zeta} \tag{6.58a}$$

for a river and

$$a_s = -\frac{\int_0^1 \hat{v}(\zeta) f_{ts}(\zeta) d\zeta}{\int_0^1 f_{tu}(\zeta) f_{ts}(\zeta) d\zeta} \tag{6.58b}$$

for a turbidity current. The minus sign on the right-hand side of (6.58a) and (6.58b) insures that a_s is positive for the physically realistic case of a vertical distribution of sand concentration that is higher toward the bed, where the secondary flow is directed inward (i.e. in the negative n direction).

Substituting (6.55a), (6.56a) and (6.57) into (6.54) and reducing with (6.45a), it is found that

$$\hat{S}_{tb} = \frac{1}{\beta}(a_b + \varphi a_s) \tag{6.59}$$

where φ is the ratio of streamwise suspended transport rate to streamwise bedload rate, as defined by (6.17).

6.6.1 River

In the case of a river carrying sand only as bedload, (6.59) reduces to

$$\hat{S}_{tb} = \frac{a_b}{\beta} \equiv A_b \tag{6.60a}$$

or in dimensioned terms,

$$S_{tb} = A_b \frac{h_o}{r_o} \tag{6.60b}$$

The formulation of (6.60) is the one inherent in Parker and Johannesson [10]. When sand is also carried as suspended load, (6.60a,b) generalize to the respective forms

$$\hat{S}_{tb} = \frac{a_b + \varphi a_s}{\beta} \equiv A \tag{6.61a}$$

$$S_{tb} = A \frac{h_o}{r_o} \tag{6.61b}$$

The formulations of (6.60) and (6.61) guarantee that the transverse bed slope is decoupled from the transverse water surface slope. In particular, according to (6.55b) and (6.56b) a_b and β are independent of S_{tt}. In addition, the forms of (6.35a) and (6.35c) are such that as must also be independent of S_{tt}.

To summarize, in a river the relation (6.40) provides a predictor of water surface slope S_{tt} (superelevation) and (6.61b) provides a predictor of bed slope S_{tb}. The two relations are decoupled from each other.

6.6.2 Channelized turbidity current

Equations (6.61a,b) also determine the transverse bed slope of a channelized turbidity current. This bed slope is no longer, however independent of the transvese interfacial slope. This can be seen by substituting (6.47a) into (6.58b) and reducing to obtain

$$a_s = a_{so} - \mathbf{Fr}_d^{-2}(\hat{S}_{tt} a_{st} + \hat{S}_{st} a_{sb}) \tag{6.62a}$$

where

$$a_{so} = -\frac{\int_0^1 f_{ts}(\zeta) \int_0^\zeta \frac{1}{\hat{\epsilon}(\zeta')} \left[-\hat{\tau}_{ni} + \int_{\zeta'}^1 f_{tu}^2(\zeta'') d\zeta'' \right] d\zeta' d\zeta}{\int_0^1 f_{tu}(\zeta) f_{ts}(\zeta) d\zeta} \tag{6.62b}$$

$$a_{st} = -\frac{\int_0^1 f_{ts}(\zeta) \int_0^\zeta \frac{1}{\hat{\epsilon}(\zeta')} \int_{\zeta'}^1 X_t(\zeta'') d\zeta'' d\zeta' d\zeta}{\int_0^1 f_{tu}(\zeta) f_{ts}(\zeta) d\zeta} \tag{6.62c}$$

and

$$a_{sb} = -\frac{\int_0^1 f_{ts}(\zeta) \int_0^\zeta \frac{1}{\hat{\epsilon}(\zeta')} \int_{\zeta'}^1 X_b(\zeta'') d\zeta'' d\zeta' d\zeta}{\int_0^1 f_{tu}(\zeta) f_{ts}(\zeta) d\zeta} \tag{6.62d}$$

Substituting (6.62a) into (6.61a), it is found that

$$\hat{S}_{tb} = \frac{a_b + \varphi \left[a_{so} - \mathbf{Fr}_d^{-2}(a_{st} \hat{S}_{tt} + a_{sb} \hat{S}_{tb}) \right]}{\beta} \tag{6.63}$$

Equations (6.47b) and (6.63) define two coupled equations for the two unknowns $\hat{S}_{tt}$ and $\hat{S}_{tb}$. They can be solved to yield

$$\hat{S}_{tt} = \frac{\gamma}{\gamma_{tt}}\mathbf{Fr}_d^2 - \frac{\gamma_{tb}}{\gamma_{tt}} \frac{a_b + \varphi\left(a_{so} - a_{st}\frac{\gamma}{\gamma_{tt}}\right)}{\beta + \varphi \mathbf{Fr}_d^{-2}\left(a_{sb} + a_{st}\frac{\gamma_{tb}}{\gamma_{tt}}\right)} \tag{6.64a}$$

$$\hat{S}_{tb} = \frac{a_b + \varphi\left(a_{so} - a_{st}\frac{\gamma}{\gamma_{tt}}\right)}{\beta + \varphi \mathbf{Fr}_d^{-2}\left(a_{sb} + a_{st}\frac{\gamma_{tb}}{\gamma_{tt}}\right)} \tag{6.64b}$$

A comparison of (6.64a,b) for the case of turbidity currents with (6.40) and (6.61a) for the case of rivers reveals an essentially different structure between the two. It thus becomes clear that (6.1a-c) are insufficient as descriptors of transverse interfacial and bed slopes in the case of subcritical turbidity currents with a sharp interface carrying substantial quantities of sand as well as mud. The processes by which transverse bed and water surface slopes are maintained for such flows are strongly coupled.

6.7 Conclusions

Strongly meandering channels on the bottom of the ocean extending for hundreds or thousands of kilometers suggest that channelized submarine turbidity currents behave in similar ways to alluvial rivers. Here the problem of steady, streamwise-uniform flow in a channel of mild, constant curvature is investigated for parallel cases of a river and a channelized turbidity current. In both cases the flow carries both sand and mud. The bed of the channel is composed of uniform sand, and this sand is carried as bedload and suspended load. The mud is transported as wash load.

Interest is focused on Froude-subcritical flows, in the standard sense for a river and in the densimetric sense for a channelized turbidity current. In the case of the turbidity current the flow is assumed to be sufficiently strongly stratified so as to have a relatively sharp interface between turbid and clear water at the top of the flow. Such an interface suppresses the entrainment of clear water from above, so allowing the flow to follow the channel for long distances.

The analysis presented here is general, but cannot be implemented in any specific case unless a turbulent closure is specified. Standard forms for predictors of transverse water surface and bed slopes are verified for the case of rivers, in which the processes maintaining the respective slopes decouple. In the case of a turbidity current carrying a substantial amount of sand with a strong vertical gradient in sand concentration, however, the standard predictors for rivers cannot be simply generalized with the use of the densimetric Froude number. In particular, the processes that maintain interfacial and bed slopes may be strongly coupled. Precise predictions based on the analysis presented here await the implementation of an appropriate closure scheme.

This research was supported by the National Science Foundation (grant no. NSF-OCE-9711431) and the STRATAFORM program of the Office of Naval Research (grant no. N/N00014-93-1-0300).

References

1. Engelund, F. (1974) Flow and bed topography in channel bends. J. Hydraul. Engrg., **100**(11), 1631-1648.
2. García, M., Parker, G. (1991) Entrainment of bed sediment into suspension. J. Hydraul. Engrg., **117**(4), 414-435.
3. García, M., Parker, G. (1989) Experiments on hydraulic jumps in turbidity currents near a canyon-fan transition. Science, **245**, 393-396.
4. Hay, A.E. (1987) Turbidity currents and submarine channel formation in Rupert inlet, British Columbia, 2, The roles of continuous and surge type flow. J. Geophys. Res., **92**, 2883-2900.
5. Imran, J., Parker, G., Pirmez, C. (1999) A nonlinear model of flow in meandering submarine and subaerial channels. J. Fluid Mech., **400**, 295-331.
6. Johannesson, H., Parker, G. (1989) Secondary flow in a mildly sinuous channel. J. Hydraul. Engrg., **115**(3), 289-308.
7. Komar, P.D. (1977) Computer simulation of turbidity current flow and the study of deep- sea channels. The Sea, 6, 603-621.
8. Mellor, G., Yamada, T. (1974) A hierarchy of turbulence closure models for planetary boundary layers. J. Atmospheric Sci., **31**, 1791-1806.
9. Parker, G., Andrews, A. (1985) Sorting of bedload sediment by flow in meander bends. Water Res. Res., **21**(9), 1361-1373.
10. Parker, G., Johannesson, H. (1989) Observations on several recent theories of resonance and overdeepening in meandering channels. In River Meandering, AGU Water Resources Monograph, Vol. 12 (S. Ikeda and G. Parker, eds.), 379-416.
11. Parker, G., Imran, J., Pirmez, C. (2000) Long profiles of leveed channels on submarine fans: a first theory. Submitted to J. Sed. Res.
12. Pirmez, C. (1994) Growth of a Submarine Meandering Channel-levee System on the Amazon Fan. PhD thesis, Columbia University, New York, USA.

Fig. 7.1. Sand bar system on the Texas coast. Bars are indicated by the patterns of wave breaking over the shallow bar depth. Three linear sand bars are visible (photo by Dag Nummedal).

wave dissipation occurs away from the shoreline, the bars provide a means of coastal protection.

Sand bars are just one example of patterns in nearshore morphology. In fact, patterns form at all scales from short ripples to large-scale variations in the bar system. Instead of blandness, it appears that nature prefers patterns, and that pattern-free morphologies are rare (hence, perhaps, of great interest). This paper addresses the general problem of pattern formation in the nearshore, away from inlets. The next section will provide a quick tour of the various scales of sediment patterns that are commonly observed. Thereafter will be a discussion of the general classes of models for pattern generation: template models, weak instabilities and strong feedback systems. Several ex-

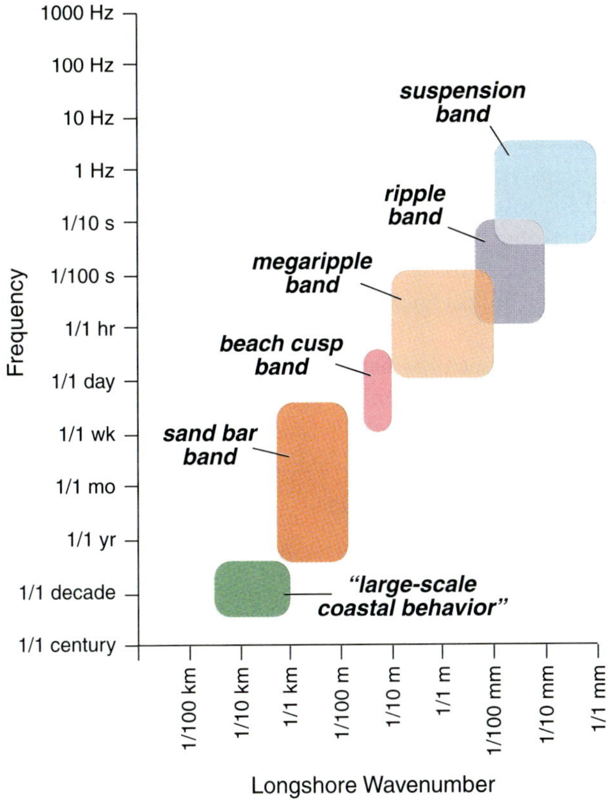

Fig. 7.2. Classes of nearshore beach features. Appropriate time and space scales are indicated on the vertical and horizontal axes, respectively. The overall slope of the line is 1 and is relatively robust to order of magnitude differences in definitions of bands.

amples of specific tests of these models will be discussed. Finally, analogy will be made to the global weather system in order to learn how to proceed.

7.3 Scales of Nearshore Morphology

The generally accepted classes of nearshore morphology are illustrated in Figure 7.2, a chart showing both the length scales (expressed as alongshore wavenumber, or inverse wavelength) and time scales (expressed as a frequency). While suspension clouds at short scales yield no morphology, the remaining blocks span sixorders of magnitude in space, from centimeters to 10's of kilometers. Time scales vary from several wave periods for ripples to decades for Large Scale Coastal Behavior (LSCB), also a span of six orders of magnitude. The slope of the line is roughly 1.0, supporting thc commonly observed relationship between space and time scales of morphological features. This slope is fairly robust to order of magnitude changes in the limits selected for different bands. Note that some conceptual morphologies such as an infinitely long linear bar do not actually occur in nature. Instead, most typically have occasional defects that provide a longshore scale.

The smallest scale patterns usually observed are wave orbital ripples. Their wavelength is typically 10 cm but depends on the wave orbital diameter (the to and fro water particle excursion taken each wave period; Figure 7.3, (Inman [14]; Komar [16]). Ripples have been studied by geologists for many years because their likely link to formative wave conditions and their common preservation in the geologic record offers the potential for ripples to act as a paleo-wavemeter (although recent observations have indicated a more complicated link between nearshore wave and current fields and the ripples and mega-ripples they produce (Traykovski et al. [25]).

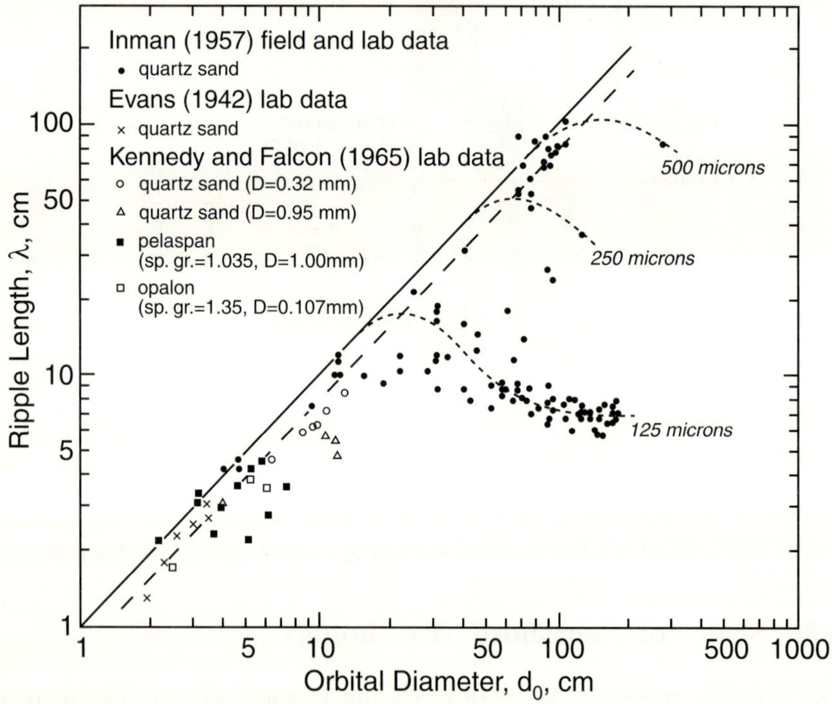

Fig. 7.3. Relationship between the wavelength of bottom ripples and the orbital diameter of the waves that formed the ripples. The unit slope for smaller waves is confused at higher wave energies (from Komar [16]).

At larger scale, beach cusps can be distinct foreshore features with longshore wavelengths of 10 to 40 m and usually a striking regularity (Figure 7.4).

Beach Cusps

Fig. 7.4. Beach cusps on the Mexican coast (from Komar [17]).

They have little offshore expression beyond a slight "ebb delta" associated with the wave backwash in each embayment.

Figure 7.2 indicates a gap between cusps and the large-scale sand bars. However, recent observations have found new morphologies in this meso-scale band. For example, short scale transverse bars, originally observed only in very quiet tide-dominated regions ([20]), have now been measured under open ocean conditions at Duck, NC, using time exposure image techniques (Figure 7.5; Konicki and Holman [18]). The features have a typical alongshore length scale of 30–100 m but a cross-shore scale that is larger (aspect ratio greater than 1). Isolated mounds of sediment have also been observed in the troughs of sand bars (Figure 7.6; Wijnberg and Holman [27]). With typical scales of 50 m, these features remain intact instead of diffusing away, and progress steadily toward the shore.

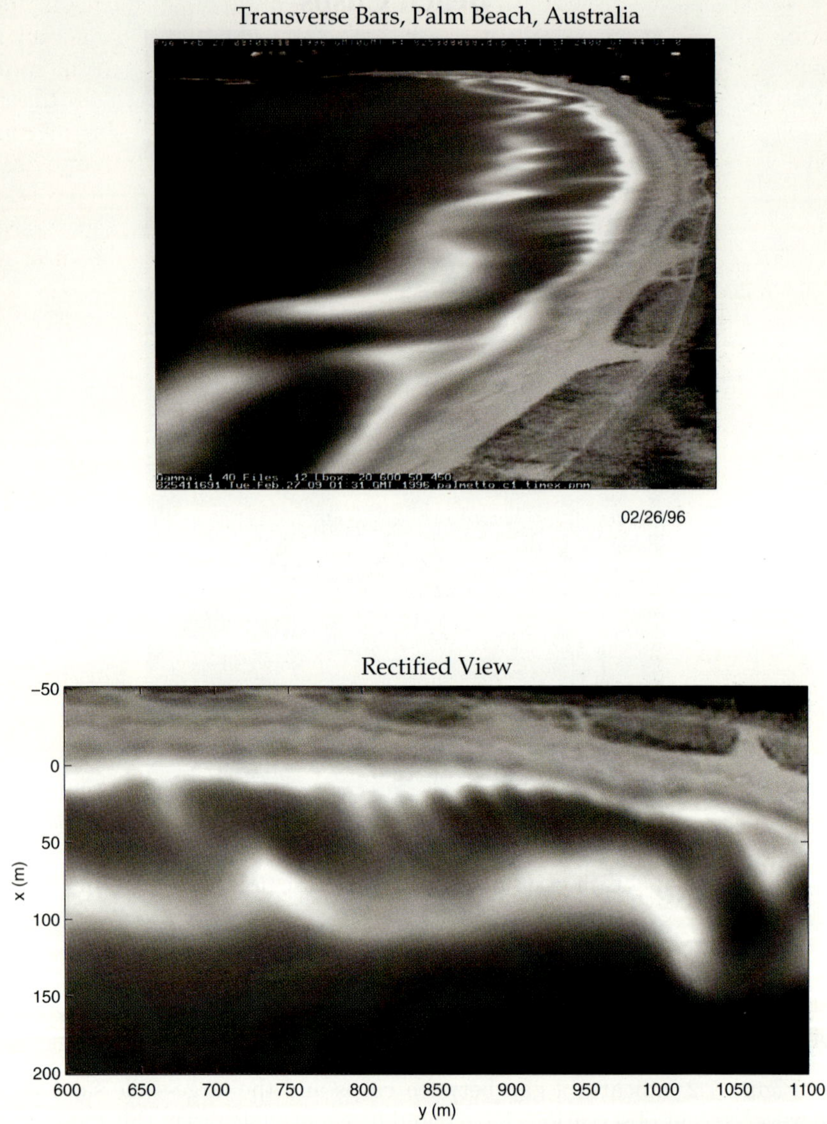

Fig. 7.5. Transverse bars on the Australian Coast, from a time exposure image showing the averaging wave breaking pattern over a ten-minute period. Similar to Figure 7.1, the white band to the left of the upper panel corresponds to wave breaking over a submerged, complex sand bar (but is smoother than Figure 7.1 due to the time averaging). A white band at the shoreline indicates the breaking of waves there (the shorebreak). The bottom panel shows a rectified, or map, view of the middle section of the beach with landward at the top and away from the camera to the left. The bands extending offshore from the shore between y = 800 and 900 are transverse bars.

At larger scales are the varying morphologies of sand bars commonly described in the literature (Figure 7.7). Sand bars are usually the dominant morphology on beaches, so their generation and response to annual variations in wave forcing have been the focus of considerable attention. Recent study of long and spatially-extensive data sets has shown surprising interannual variability of these bar systems, a variability now know as Large Scale Coastal Behavior. Thus, observations of the same system over many years shows very different sand bar configurations both spatially and temporarily despite apparently identical offshore (wave) forcing conditions (Plant et al. [22]; Wijnberg and Terwindt [28]).

Fig. 7.6. Time exposure image of the shore and sand bar at Duck, North Carolina, USA. The figure shows an isolated blob of breaking between the shore and the sand bar that corresponds to a localized shallow region. Surprisingly, this feature stayed intact for two more weeks, propagating toward the shore, and is called a Shoreward Propagating Accretionary Wave, or SPAW. The full sequence, from formation to absorption by the beach, took approximately three weeks.

Fig. 7.7. Examples of sand bar morphologies observed in two years of time exposure images from Duck, NC. Morphologies vary from linear (bottom right), to crescentic and regular (middle left) to complex (top right).

7.4 Mechanisms of Pattern Formation

Mechanisms for the generation of patterns from an initially monotonic beach profile can be considered under three categories: template mechanisms; weak instability mechanisms, or strong feedback mechanisms. In this section, we will discuss the roots of this distinction and some broad characteristics of each class. Then in the next section, we will quickly describe several model tests, particularly from the template class.

The distinction between classes of mechanisms can be explained in terms of a general organizational figure for nearshore processes (Figure 7.8). The nearshore system consists of both fluid and sediment components. At its broadest level, nearshore dynamics is the study of the evolution of a wave field as it progresses over a shoaling and potentially complex coastal bathymetry, $h(x, y)$, and the slower evolution of the bathymetry to those overlying flows. The study of morphology generation is, by definition, the study of the bathymetric response. However, knowledge of the fluid component is still required.

The upper panel in Figure 7.8 provides a cartoon representation of both the shoaling wave transformation and of the bathymetric response. The lower panel illustrates the relationships in a block format. Thus, fluid transformation processes are simply represented as a box called "fluid dynamics" that depends on bathymetry only as a boundary condition (noted for simplicity as a function of the cross-shore location only, but really a function of both cross-shore and longshore dimensions). Bathymetric response is accomplished by the dashed feedback loop by which the processes of sediment transport yield a rate of change of bathymetry.

The distinction between the three mechanisms of morphology generation depends on the description of the feedback loop. For template mechanisms the loop is severed so that the boundary condition, $h(x)$, for the fluid dynamics is adequately described by its initial state, $h_0(x)$. This may be because Δh is either small or has simply been ignored. As an example, the rearrangement by wave motion of a small number of grains of sand, initially sprinkled uniformly across a cement beach, into patterns would be a template mechanism since the impact of the sediment transport on the bathymetry would always be infinitesimal. For weak instabilities, the feedback loop exists (and must exist) but the impact of the changing bathymetry on fluid motions can be described in simple analytical (perturbation) terms. The spontaneous development of wave orbital ripples is likely a weak instability mechanism. Strong feedback means that bathymetric changes are sufficiently large that consequences to fluid processes cannot be adequately described by perturbation techniques.

Note that feedback may also be negative, such that a perturbation to bathymetry will be damped out in time. Since no morphology can be generated with negative feedback, only cases of positive or zero feedback are considered in this discussion.

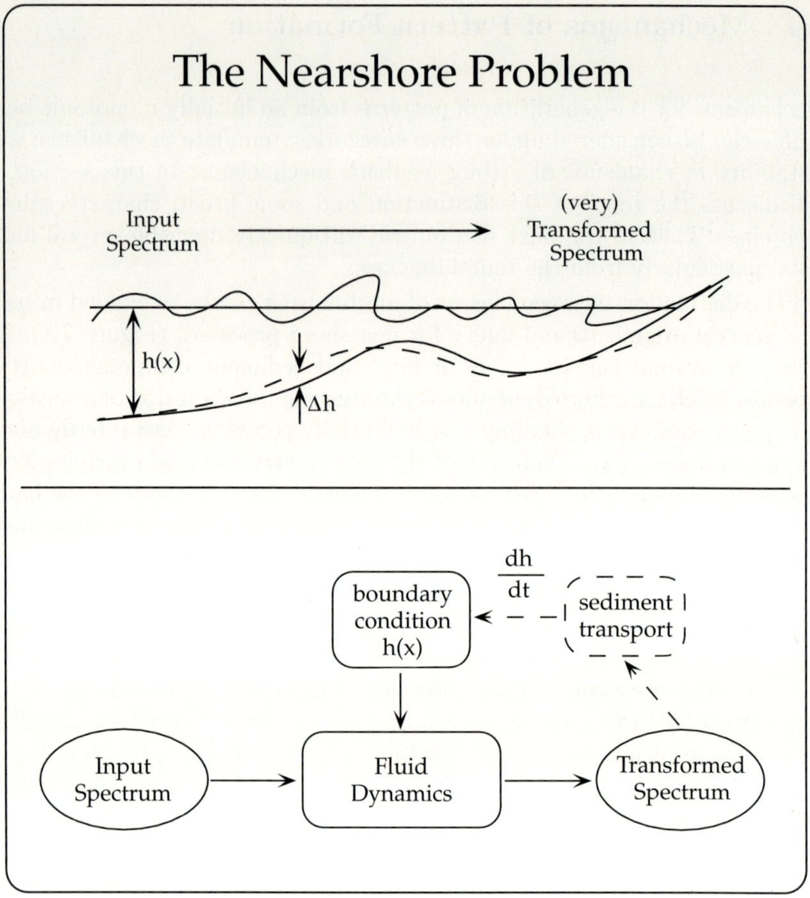

Fig. 7.8. Schematized figure illustrating the basic nature of the nearshore problem. Upper panel illustrates transformation of waves due to the shoaling bathymetry and the corresponding evolution of the bathymetry. Lower panel shows the same idea in a block diagram. The key to complex morphologies lies in the feedback loop, labeled sediment transport, that allows the fluids to affect the bottom boundary, which in turn affects the fluid processes.

The distinction between the different classes of pattern formation also depends on the time scale of interest. For instantaneous predictions (now-casts) of system behavior, changes in bathymetry, Δh, are zero. Thus the dominant physics is always either template (if the pattern arises from enforced scales in the fluid) or weak instability (if scales arise spontaneously). Similarly, a short-term prediction could be dynamically defined as one for which bathymetric response is small and could be handled either through analytic perturbation or time-stepping numerical model. A long-term prediction would be one for which the evolution of the bathymetry is "large". The definition of "large" is again dynamical, and implies that the accumulative errors of perturbations or time-stepping become unacceptable. In the literature for nonlinear dynamical systems, this concept is captured by the idea of a prediction horizon, the time scale longer than which two realizations of nearly identical initial conditions have diverged unacceptably. Thus, feedback is considered to be "strong" if time scales of interest are beyond the prediction horizon. It should be kept in mind that this time scale could be as short as hours for rapidly evolving beach cusps, or perhaps a month for a typical sand bar.

7.5 Template Mechanisms

The defining characteristic of template mechanism is that the initial bathymetry, $h_0(x)$, is an adequate description of the boundary condition for waves and flows for all time. Thus, a simplistic equation for the evolution of the morphology might be:

$$\frac{\partial h(x,\tau)}{\partial t} = f\left(\text{fluids }(h_0,\tau), \text{sediments}\right) \tag{7.1}$$

where f is a function describing the dependence on fluid dynamics and, potentially, on the nature of the sediments through grain size distributions or other qualities. The fluid dependencies, denoted "fluids" for simplicity, include all external wave (and wind) forcing, specified perhaps by bulk parameters such as significant wave height, direction and period or perhaps through a deep water directional spectrum, and the evolution of this forcing through shoaling transformation across the bathymetry, h_0. The external components of waves may vary slowly with time (τ), such as for storms, but the boundary condition for flows does not. If h were represented by a vector of values at locations x_i, then the above equation would represent a coupled set of differential equations, one for each i.

Inherent in the definition of the template mechanism is the fact that the scale of the morphologic features is determined by scales within the flow. Thus, Figure 7.2 can be overlain with a corresponding figure of the time and space scales of dominant flows within the nearshore (Figure 7.9). So sand bars, for example, are commonly associated with infragravity waves due to their similar cross-shore and longshore length scales.

Template models were most popular in the 1960's through the 1980's because they provided elegant, testable models. This will be illustrated by three examples of specific template models; for the generation of sand bars, beach cusps and rip currents.

7.5.1 Sand bar generation by standing infragravity waves

Miche [19] suggested that waves, propagating toward a beach, can be decomposed into a component that is progressive and decays to zero at the shoreline

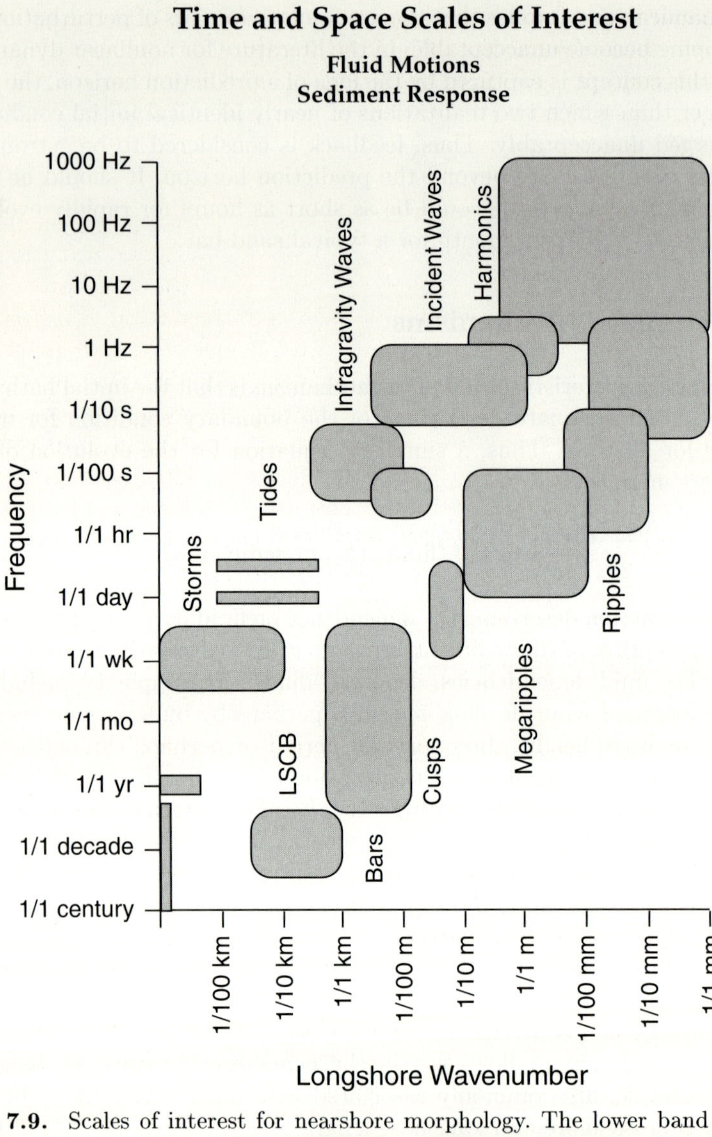

Fig. 7.9. Scales of interest for nearshore morphology. The lower band of classes are the observed classes of morphology (identical to Figure 7.2). The upper band corresponds to dominant bands of fluid motion. Template models of morphology generation explain length scales as resulting from scales of dominant fluid motions. Hence, the physics is isolated to vertical bands in this figure.

and a component that is standing (perfectly reflected at the shoreline). The reflected component has a pattern of nodes and anti-nodes at cross-shore locations that depend on the period of the wave. Sand, sprinkled loosely on the bottom, will tend to accumulate under either nodes or anti-nodes (depending on the dominant type of sediment transport) of these standing waves forming the initial pattern of a sand bar (Bowen [5]; Short [24]).

For a plane beach (a simple uniform slope offshore) of slope β, the predicted bar location, x_b, would be determined by the relation

$$\frac{(2\pi)^2 f^2 x_b}{g\beta} = 4.5 \tag{7.2}$$

where f is the wave frequency (reciprocal of the wave period) and g is the acceleration of gravity. In some cases, the cross-shore scaling of bars can be short, consistent with reflection of ocean waves (Carter et al. [6]). However, sand bars are typically located about 100 m offshore and beach slopes are roughly 0.02. Thus, the required wave period would be of order 66 seconds, greatly exceeding normal ocean values (O(10 s)). Motions at these longer periods are given the special name infragravity waves, and have received considerable attention due to their possible relationship to bar generation (an alternate model for the generation of long spatial patterns by waves has been proposed by Boczar-karakiewicz and Davidson-Arnott [3] based on a narrow-band instability mechanism, but applies only to very narrow-banded wave conditions).

Specific tests of this hypothesis (in terms of cause and effect) have been surprisingly difficult. However, two minimal requirements for this model to be relevant to natural beaches are:

1. There must be significant infragravity energy present for bars to form, and
2. This energy must be in the form of a dominant spectral peak.

If the latter criterion is not satisfied, then the bar development at a location predicted for a single frequency must be averaged over locations appropriate to the entire spectrum and no single bar will dominate. While the definition of "dominant" in point 2 depends on a full sediment transport model, a minimal requirement is that the spectral peak should be at least statistically significant.

On most beaches around the world, criteria 1 is satisfied, at least during storm conditions (Guza and Thornton [9]; Holland and Holman [11]; Holland and Holman [10]; Howd et al. [6]). Criterion 2 was addressed by [11] using wave runup (the seaward limit of the moving water against the beach face) collected over the 16-day field Delilah field experiment at Duck, NC, in 1990. They found no significant differences between the observed runup spectra and that that would be associated with a Gaussian noise process. In other words, infragravity motions clearly influence nearshore sediment transport, but cannot provide a template for bar formation at a specific location.

A second popular template mechanism concerns the generation of rhythmic beach cusps (Figure 7.4). Guza and Inman [8] suggested that cusps were carved out by edge waves (alongshore propagating waves trapped to the shore

bottom ripples are most likely caused by such an instability. Since the influence of growing ripples on the flow is limited to a small fraction of the water column, the strength of the feedback remains small and instability models are useful. Instability models have also been used to explain transverse sand bars that are sometimes observed in mild wave environments(Barcilon and Lau [2]; Falques [7]). This case is similar to bedform generation in rivers except that the primary flow comes from longshore currents and, since the bottom slopes offshore, the perturbation to the current takes the form of an arrested edge wave. However, a recent paper by Konicki and Holman [18] finds no evidence to support existing formulations of this simple mechanism on an exposed coastal site.

While the above discussion is founded on the use of analytic approaches, it equally applies to the numerical modeling of nearshore morphology. At each time step, the fluid dynamics of the system are computed using the most recent bathymetry as the boundary condition, $h_0(x)$. From the resulting sediment transport, a small increment to the bathymetry, $\epsilon(x)$, is found and the prior bathymetry replaced by $h_0 + \epsilon$. With this approach, the evolution of the nearshore bathymetry can be modeled beyond the time when ϵ has become a large departure from the original bathymetry.

7.7 Strong Feedback Systems

It would seem that the above approach of time stepping would allow modeling of morphologic evolution over any length of time. However, research over the last twenty years on nonlinear dynamical systems has shown this to be untrue Abarbanel [1]. Two nearly identical initial bathymetries, subjected to identical forcing, will slowly diverge until, after a length of time, they become uncorrelated. This concept is captured by the idea of a prediction horizon in time. For shorter time scales, time stepping models can be successful. Beyond the prediction horizon, even perfect models, applied to initial data that has some noise, will produce results that have no relationship to reality.

An excellent analogy exists with the global weather and climate system. The processes by which the weather changes occur primarily at small scale with physics that is fairly well known. Thus it would seem that models could be integrated over long times to predict climate. However there are many feedbacks and the system has many patterns and complexity at all scales (Figure 7.11).

Weather predictions for time scales up to one week, computed by integration of numerical models, have become increasingly accurate. However, accuracy degrades rapidly for longer predictions and even perfect physics and excellent initial data will not likely allow accurate predictions beyond a few weeks. Yet it is the longer-term climate that can most dramatically affect society. For example, knowledge of impending droughts or El Niño events would be invaluable for societal planning.

Fig. 7.11. Atmospheric cloud patterns around the Southern Ocean.

One way to proceed would be to take the best available small-scale model of global weather and integrate it forward many times, in a Monte Carlo sense. While each realization would not likely match the future behavior of the true system, it is hoped that their ensemble statistics would provide a statistical prediction of long term behavior. However, the quality of the prediction will likely be limited by the quality of the physics in the model and the required computation time would likely be prohibitive, particularly if all of the relevant, detailed physics were included.

Surprisingly, the best predictors of ENSO (El Niño, Southern Oscillation, a major interannual variation in global climate) events are not based on integration of the full weather system forward in time, but instead on observation of a few empirical indices such as the ENSO index (Philander [21]), the difference in sea level at two sites spanning the equatorial Pacific. Identification of the important aspects of the key mechanisms that determine variability at long scales of interest is a primary scientific challenge.

Thus, there are several important characteristics of the global climate and weather systems that can be compared to the nearshore system:

- The global atmosphere is a complex system with patterns at many scales
- Physics at smaller scales (weather) are relatively well known
- Primary societal interest is in climate, rather than weather
- Substantial interannual variability exists, even in the absence of interannual forcing
- Knowledge of links between near term and climate is primitive.

The list is nearly the same as would be generated for the nearshore system. The nearshore is a complex system involving interactions between the fluid and sediment components at many scales. Changes occur due to short scale physics (wave dynamics that are reasonably well known and coupled with sediment transport that is less well understood). However, societal interest is centered on "beach climate" (multi-year trends for which society can plan).

As with the weather-climate example, it is likely that simple forward integration of even perfect equations for nearshore evolution will begin to fail after sufficiently long times (we currently have no way to estimate the value of this prediction horizon time scale, but it might be several months for many beaches). Moreover, at these longer scales, surprising variability exists. Wijnberg and Terwindt [28] have shown that sand bars on the Dutch Coast are actually offshore progressive, such that new bars are generated at the coast, slowly move offshore, then decay over a period of years. The time scale of this cycle jumps from about 4.5 years in the Southern Dutch Coast to 14 years in the north for no known reason. Plant et al. [22] observed a similar pattern of offshore bar movement at Duck, NC, although with less regularity and with substantial differences on either side of a research pier located at the site. These interannual signals apparently cannot be explained by any known interannual variations in system forcing. In addition, as for the global atmosphere (Figure 7.11), the nearshore bar system can be complex (Figure 7.12), particularly compared to patterns that might be expected from the wave forcing.

The problem then is to find an efficient set of links between short term processes and resulting long term variability. We may adopt a hypothesis that, of the many feedback loops and interactions within the nearshore system, some may provide the dominant influence in determining the long term variability of the system. Many other processes may then be neglected by comparison (although their short term influence may be substantial).

There are several approaches available in this search for simplification. The first involves integration of full models forward in time to determine behavior and to understand sensitivity of output to component processes. The idea, again, is to search for key processes from a full dynamics framework.

Fig. 7.12. Rectified (map) time exposure image of sand bars at Duck, NC. The pier at Duck is visible on the left (y = 500 m) while the sandy beach lies in a band across the middle of the image with seawards to the top of the figure. The bands of white indicate the most complex sand bar system the author has seen.

A second approach involve self-organization models, wherein a selected subset of the physics are implemented in terms of a series of rule that are stepped forward to generate morphology (Werner and Fink [26]). While the representation of the physics are often crude, these models reveal the potential for even very simple coupled systems to develop complex and highly variable patterns. These models also run quickly on computers since only a few processes are included. The contribution of these models is to allow empirical exploration of the minimum physics required to explain observed patterns. A challenge for this approach is to provide a more formal connection between the selected set of rules and accepted laws of physics.

The third approach to understand long term system variability is top down modeling. In this approach, phenomenology observed in long term data sets are used to constrain the physics required for modeling. For example, the offshore progression of the Dutch sand bars requires the addition of new physics that are not incorporated in current models. The discovery of these new phenomena has only resulted by observation, particularly observation over long periods of time and at types of locations that have not already been extensively measured.

7.8 Discussion

An overall concern with modeling any complex system is the development of an appropriate set of measurements of system state and of the ability of a model to predict evolution in state. Within this paper, I have side-stepped this question by simply referring to the state of the nearshore system as $h(x)$, with no detail on whether h is represented by a single function, the superposition of many functions, or by a discretized representation with some sampling interval, Δx. Moreover, I omitted any variability in the longshore direction, obviously a gross simplification (Figure 7.12). As the amount of information required to specify system state increases, so also does the complication in specifying the success of any model.

The search for simplicity revolves around the search for dominant processes, allowing many others to be neglected. In the nearshore, wave breaking stands out as such a process. When waves break, the momentum that they carry is transferred to the water column to drive flows such as undertow. The strength of these forces are of order 10^2 to 10^3 dynes/cm^2, among some of the strongest in the ocean (by contrast, the wind stresses that drive the Gulf Stream are about two orders of magnitude smaller). Thus, the presence or absence of wave breaking over a sand bar provides a huge feedback between an incipient bar and the flows that may have caused the bar. Investigation of the processes and consequences of wave breaking in the nearshore are a very important research topic.

The term "strong feedback system" is awkward in some ways. The defining aspect is really one of time, and implies that the time scale of interest is beyond the prediction horizon. This time scale would be short for systems with strong nonlinearities or strong feedbacks. However, even a weakly nonlinear system can be unpredictable after a long enough period of time and thus would be considered a strong feedback system within this definition.

7.9 Conclusions

The form taken by natural sandy beaches has patterns at a wide range of scales. Most early models for the generation of these forms sought aspects of the wave forcing that had the same physical scales, such that these fluid motions served as a template for the resulting sand bars and other morphologies. However, recent tests of a number of these models have found no evidence of dominant scales being introduced by fluid motions. Instead, the forcing of the system appears to be bland.

Patterns may also be generated spontaneously through instabilities in which there is positive feedback between perturbations in the bathymetry and the overlying fluid motions. In the case where these perturbations are small either by assumption or through other physical mechanisms, a range of analytical tools are available to study the growth of features. These can also be treated numerically. Larger departures from an initial state can be treated by time stepping models, such that the perturbations are always only a small departure from an updated initial state.

Most societal interest in understanding the evolution of sandy beach environments lies at the long time scales of coastal planning, well beyond the time scale for which time stepping integration can provide reliable results. Long term field observations at a few ocean sites have shown that unexpected phenomena can arise. Progress will require ongoing sampling programs at different sites to determine the natural range of behavior, coupled with modeling efforts to determine the simplest level of physics that can explain these long term phenomena. In this way, studies in the nearshore are similar to efforts to explain long term global climate variability such as El Niño. The relative roles of bottom up physics, self-organization and top-down models toward a final solution is unknown.

References

1. Abarbanel, H. D. I. (1995) Analysis of observed chaotic data. Springer-Verlag, New York, 272 pp
2. Barcilon, A. I., Lau, J. P. (1973) A model for the formation of transverse bars. J. Geophys. Res. **78**, 2656–2664
3. Boczar-karakiewicz, B., Davidson-Arnott, R. G. D. (1987) Nearshore bar formation by non-linear wave processes – A comparison of model results and field data. Marine Geology **77**, 287–304
4. Bowen, A. J. (1969) Rip currents, 1. Theoretical investigations. J. Geophys. Res. **74**, 5467–5478
5. Bowen, A. J. (1980) Simple models of nearshore sedimentation; beach profiles and longshore bars. In: McCann S.B. (ed) The Coastline of Canada, Geological Survey of Canada, 1–11
6. Carter, T. G., Liu, P. L., Mei, C. C. (1973) Mass transport by waves and offshore sand bedforms. Journal of the Waterways, Harbors and Coastal Engineering Division, WW2, 165–184
7. Falques, A. (1991) A note on the Barcilon and Lau model for transverse bars. Review de Geofisica **47**, 191–195
8. Guza, R. T., Inman, D. L. (1975) Edge waves and beach cusps. J. Geophys. Res. **80**, 2997–3012
9. Guza, R. T., Thornton, E. B. (1982) Swash oscillations on a natural beach. J. Geophys. Res. **87**, 483–491
10. Holland, K. T., Holman, R. A. (1996) Field observations of beach cusps and swash motions. Mar. Geol. **134**, 77–93
11. Holland, K. T., Holman, R. A. (1999) Wavenumber-frequency structure of infragravity swash motions. J. Geophys. Res. **106** (C6), 13479–13488
12. Holman, R. A., Sallenger, A. H., Jr. (1985) Setup and swash on a natural beach. J. Geophys. Res. **90** (C1), 945–953
13. Howd, P. A., Oltman-Shay, J, Holman, R. A. (1991) Wave variance partitioning in thc trough of a barred beach. J. Geophys. Res. **96** (C7), 12,781–12,795

14. Inman, D. L. (1957) Wave-generated ripples in nearshore sands. U.S. Army Corps of Engineers, Beach Erosion Board, 66 pp
15. Kennedy, J. F. (1963) The mechanics of dunes and antidunes in erodible-bed channels. J. Fluid Mech. **16**, 521–544
16. Komar, P. D. (1974) Oscillatory ripple marks and the evaluation of ancient wave conditions and environments. J. Sediment. Petrol. **44**, 169–180
17. Komar, P. D. (1976) Beach Processes and Sedimentation. Prentice-Hall, Englewood Cliffs, NJ, 429 pp
18. Konicki, K. M., Holman R. A. (in review) Transverse sand bars in the nearshore. Mar. Geol.
19. Miche, R. (1951) Le pouvoir réfléchissant desouvrages maritimes exposés à l'action de la houle. Ann. Ponts Chaussees **121**, 285–319
20. Neidoroda, A. W., Tanner, W. F. (1970) Preliminary study of transverse bars. Mar. Geol. **9**, 41–62
21. Philander, S. G. (1990) El Nino, La Nina and the Southern Oscillation. Academic Press, San Diego
22. Plant, N. G., Holman, R. A., Freilich, M. H. (in press): A simple model for interannual sand bar behavior. J. Geophys. Res.
23. Ranasinghe, R., Symonds, G., Holman, R. A. (in review) Rip spacing and persistence on a swell dominated beach. J Geophys. Res.
24. Short, A. D. (1975) Multiple offshore bars and standing waves. J. Geophys. Res. **80**, 3838–3840
25. Traykovski, P., Hay, A. E., Irish, J. D., Lynch, J. F. (1999) Geometry, migration and evolution of wave orbital ripples at LEO-15. J. Geophys. Res. **104** (C1), 1505–1524
26. Werner, B. T., Fink, T.M. (1993) Beach cusps as self-organized patterns. Science **260**, 968–971
27. Wijnberg, K. M., Holman, R. A. (in review) Shoreward propagating accretionary waves in the nearshore. J. Geophys. Res.
28. Wijnberg, K. M., Terwindt, J. H. J. (1995) Extracting decadal morphological behavior from high-resolution, long-term bathymetric surveys along the Holland coast using eigenfunction analysis. Mar. Geol. **126**, 301–330

8 Long-Term Morphological Prediction

Huib De Vriend[1,2]

[1] Fac. Civil Engineering & Geosciences, Delft University of Technology, POB 5048, 2600 GA, Delft, NL

[2] Dept. Civil Engineering & Management, University of Twente, POB 217, 7500 AE Enschede, NL

8.1 Abstract

Long-term morphological changes of alluvial rivers, estuaries, coasts and shelf seas are often difficult to predict, not only because part of the external forcing (e.g. the weather) is only predictable in a statistical sense, but also because the non-linearity of the morphodynamic processes may give rise to inherent predictability limits, which are associated with multiple equilibrium states, or even with deterministic chaos. The statistical properties of these systems are only predictable at scales much larger than those of the underlying processes or forcing factors. Hence there is a need for predictive models that are defined at the scale level of interest, rather than at the much smaller scale of the underlying processes. This paper discusses possibilities to formulate and utilise such aggregated models.

This emphasizes the importance of behaviour analyses of idealised models as a method of investigation, in addition to (1) field measurements and monitoring, (2) laboratory experiments and (3) numerical modelling. The paper gives an overview of recent work on such behaviour analyses and discusses the possibilities of further integration with the other three methods.

A third complication concerns parameter setting, calibration and sensitivity analysis. In general, a model may lose its predictive capability (other than straightforward extrapolation from the past) if too much of its essential dynamics is parameterised and calibrated against data from the past. In other words: a model can be over-calibrated and thus be deprived of its predictive capability. The awareness of this possibility is virtually non-existent in morphological modelling, let alone the notion of optimal calibration. In this paper, we discuss some examples of sensitivity analysis and over-calibration of morphological models.

8.2 Introduction

In active alluvial systems, such as rivers, estuaries, coasts and shallow shelf seas, sediment is moved around by the water. This leads initially to morphological changes at the spatial and temporal scales of the water and sediment motion, which in their turn influence the flow and sediment transport processes. This dynamic coupling between the water/sediment-motion and the bed topography may lead to morphological changes at much larger spatial and temporal scales than initially imposed by the hydrodynamic and sediment transport processes.

The ability to accurately predict these large-scale changes, whether natural or due to human interference, is increasingly important, not only because the systems we are dealing with are under increasing pressure (multiple usage for a variety of competing functions, loading the system up to its carrying capacity or beyond), but also because sustainable development has become a central issue in the management of these systems. A special aspect of sustainability is the long-term evolution of morphology-related system-properties, such as the flood and ice conveyance capacity of a river, the navigable depth of shipping channels, the resilience of a dune coast acting as a sea defence, or the stability of the morphological state of a tidal lagoon (will it remain shallow with a high percentage of intertidal area, or will it flip over to deep with hardly any intertidal flats?). Risk analyses concerning these properties put special demands on the long-term predictive capability of morphological models.

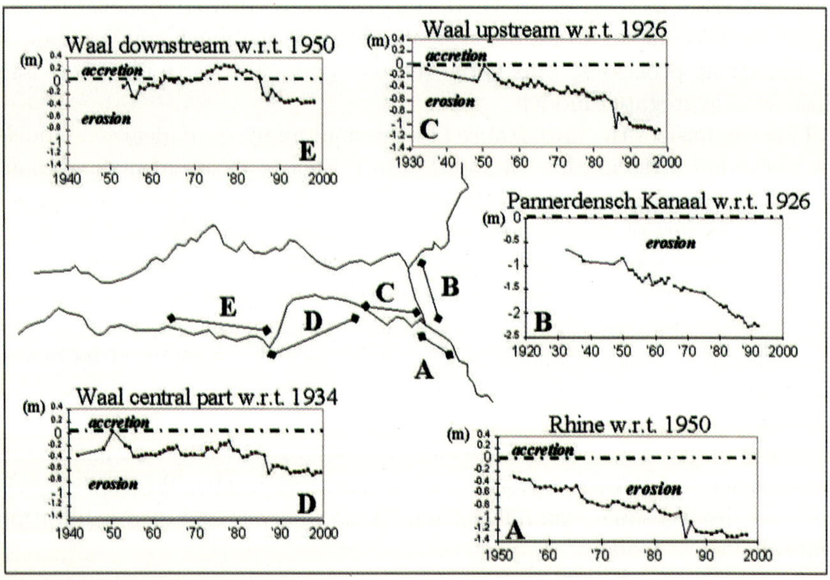

Fig. 8.1. Long-term response of the Dutch Rhine branches to river training works and other interferences in the late 19th and early 20th century [courtesy Rijkswaterstaat; also see Visser et al. [44]].

A practical example of a long-term response to human interference concerns the Dutch Rhine branches (Figure 8.1; also see Visser et al. [44]). In the late 19th and early 20th century, significant training and normalisation works have been carried out, resulting in an overall narrowing of the river. River narrowing is known to lead to a decrease of the overall bed slope (e.g. Jansen [18]). Other interferences, such as sand mining in the lower reaches and sand loss due to subsidence by the collapse of old mines below the German part of the Rhine, had a similar effect. As a consequence, the bed in the upper reaches of the Dutch Rhine branches has eroded up to 2 m over the last half century[1]. Clearly, this leads to problems in the maintenance of river works, bridge piers, etc.

Another example of man-induced long-term changes is the Western Scheldt, The Netherlands, where fourty years of channel deepening and maintenance have not only led to a hydrodynamic response (a significant increase of the high-tide level near Antwerp), but also to a morphological one, viz. a dramatic reduction of the secondary channels and the intertidal area (Vron et al. [49]). This morphological reponse means the loss of much of the natural multi-channel character of the estuary.

A third example concerns the Bay of Mont-Saint-Michel, France. In the past, the Mont used to be surrounded by regularly flooded mudflats, but now these have turned into marshland, presumably due to the polderisation of the old marshes in the beginning of the twentieth century (e.g. Migniot [23]).

An example of long-term natural changes is described by Isreael and Dunbergen [17] and earlier, for another case, by Oost [27]. It concerns the quasi-cyclic migration of the main channels in a tidal inlet, with a period of approximately 50 years. Note that the main driving force of this system is the tidal motion, with a cycle of half a day!

Another example of natural long-term behaviour is the evolution of nearshore bars, which are generated near the beach, gradually migrate offshore and ultimately vanish, with typical lifetimes of several years to decades (Wijnberg and Terwindt [50], Holman [15]). The main driving force of these features is probably formed by moderate storm events, occurring several times per year.

Apparently, these morphodynamic systems can exhibit a significant long-term response, at time scales far beyond the basic frequency of their forcing.

In summary, there is definitely a practical need to predict long-term morphological changes in rivers, estuaries, coasts and shelf seas. This paper is meant to contribute to the development of a research strategy that must enable us to properly deal with such long-term predictions.

[1] This erosion is now called 'autonomous degradation', probably because people have lost track of its origin. Recent investigations (Visser et al. [44]) have shown, however, that it can be contributed entirely to recorded human interferences.

8.3 Definition of Long-Term/Large-Scale

The terms *long-term* and *large-scale* need further definition, in order to be practically operational. There seems to be a natural tendency towards absolute definitions (longer than so many years, larger than so many kilometres). Such definitions can be used in a particular situation, but they are not generally applicable.

A more generic approach starts from a scale level that is related to the morphological system, such as the inherent time scale, $T_{process}$, of the predominant underlying process (e.g. wave period, tidal period, settling time of suspended sediment), or the time scale, $T_{forcing}$, of the external forcing (recurrence interval of flood waves, storms, surges, etc.), or the inherent time scale, T_{morph}, of the predominant morphological changes[2] (lifetime of a bedform, period of a channel migration cycle, lifetime of a nearshore bar or a sandwave). The qualification *long-term* in that case refers to time scales much larger than these, so $T \gg T_{process}$, or $T \gg T_{forcing}$, or $T \gg T_{morph}$. Depending on the situation, these definitions are close together (in the swash zone, for instance), or they refer to entirely different scale levels (on the lower shoreface, for instance, or on the continental shelf, where the transport rate is usually very small). Henceforth, we will use the forcing-related definition, unless stated otherwise.

In principle, similar definitions could be made for the spatial scale. We assume, however, a certain degree of coupling between spatial and temporal scale levels (also see Figure 8.2), arguing that larger-scale changes generally take longer to develop. As a consequence, there is a certain band in the space/time-scale domain in which spatial and temporal processes are dynamically interacting (e.g. in the form of migrating spatial patterns). Outside this band, we have either slowly varying spatial details, or rapidly varying large-scale patterns ('unrest'). If existent, at all, these phenomena are not likely to be of interest to the processes within the dynamic band. Note that this concept is not necessarily in contradiction with the aforementioned non-linear scale interactions, which occur primarily between either spatial or temporal scales.

8.4 Free and Forced Behaviour

If we consider morphology as the manifestation of a non-linear dynamical system, we can use systems theory to explain morphological behaviour. One of the concepts encountered there is the distinction between forced and free

[2] In general, the morphological time scale is proportional to the product of two length scales, characterising the morphology of interest, and inversely proportional to the volumetric sediment transport rate per unit width. In the case of bedform propagation, for instance, the two length scales characterise the length and the height of the bedforms, respectively.

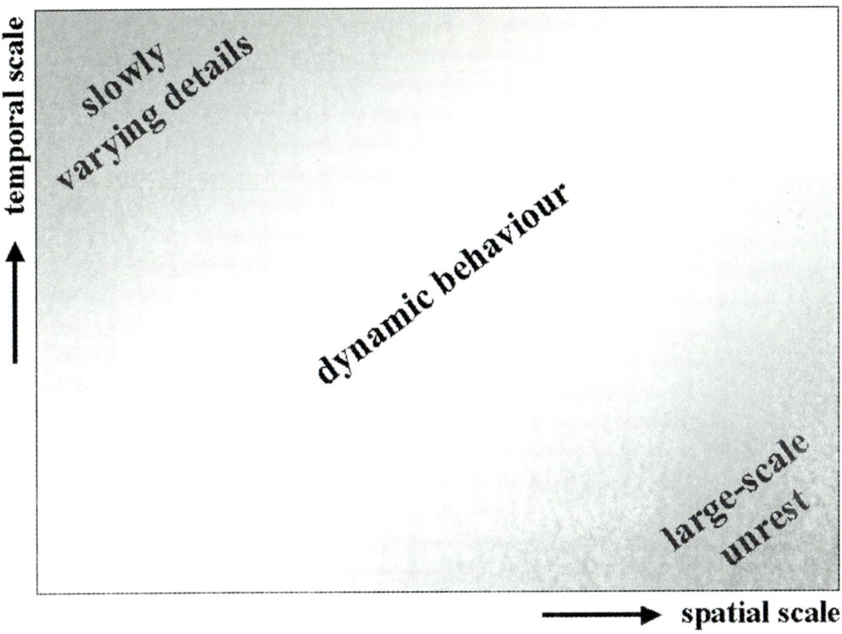

Fig. 8.2. Coupling of spatial and temporal scale.

behaviour. In the former case, the variations in the output can be attributed to the variations in the forcing, in the latter case there is no correlation between the two.

There are many examples of free morphological behaviour (also see De Vriend [7,8]):

- in the case of river morphology: bed-ripples and dunes, alternate bars, meandering and braiding, channel pattern evolution;
- in the case of estuarine morphology: bed-ripples and dunes, channel/shoal patterns, meandering (see Figure 8.3);
- in the case of barrier-island coasts: bed-ripples and dunes, channel/shoal patterns (on the outer outer delta and in the basin), longshore migration of islands and inlets;
- in the case of sandy beach barrier coasts: bed-ripples, beach cusps, near-shore bars, rip-cells, shoreline waves, shoreface-connected ridges;
- in the case of shallow shelf seas: bed-ripples and megaripples, sandwaves, tidal sandbanks.

Many such phenomena are described in papers to the RCEM-symposium (Seminara [35]). None of them can be attributed to variations in the forcing at the same spatial and temporal scales. A global energy input (i.e. without much spatial or temporal structure), via currents, waves or tides, is sufficient to trigger their formation. Rather are these phenomena manifestations of the system's internal dynamics.

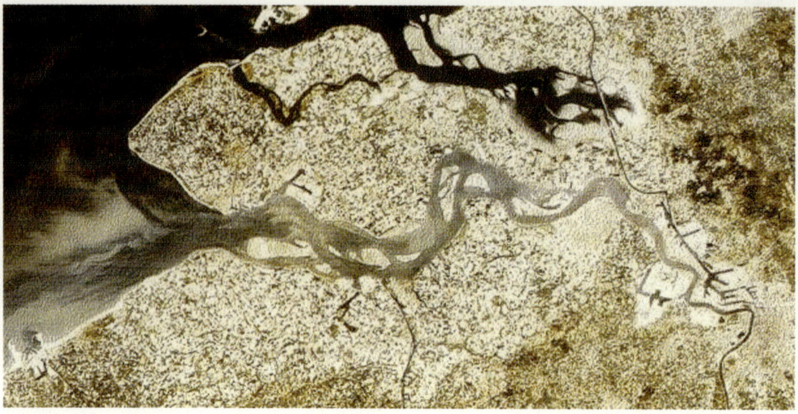

Fig. 8.3. Morphology of the Western Scheldt estuary, The Netherlands. At least three morphological scales are visible: the meso-scale (in channel morphology), the macro-scale (channel/shoal pattern) and the megascale (estuary as a whole) [courtesy Rijkswaterstaat].

Examples of forced behaviour are:

- in the case of rivers: response to dredging and dumping of sediment, response to hydraulic structures, training works, nature restoration works, etc.;
- in the case of estuaries: response to channel dredging and maintenance, sand mining, land reclamation, polder abandonment, etc.;
- in the case of coastal lagoons: response to channel deepening and/or fixation (e.g. by jetties or guiding walls), sand mining, land reclamation or abandonment, relative sealevel rise, etc.;
- in the case of beach barrier coasts: response to shore nourishments, jetty construction, detached breakwaters, changes in wave climate (e.g. due to climatic events like El Niño; Simpson [37]), relative sealevel rise, etc.;
- in the case of shelf seas: response to channel dredging and maintenance, sand mining, construction of artificial islands, etc.

The temporal and spatial scales of this type of response are associated to those of the forcing.

A mixed form is constituted by the so-called switch-on phenomena, i.e. the dynamic response to an initial perturbation, of which the spatial scales are correlated to those of the initial perturbation, whereas the time scales are inherent to the system. A possible example, which needs further investigation, is sand mining in shelf seas: certain dimensions and orientations of the borrow pit may trigger the formation mechanism of tidal sandbanks (e.g. Hulscher [16]). Moreover, certain types of forcing may lead to a resonant response of the system, sometimes in significantly larger areas than where the forcing applies.

8.5 State of the Art in Morphological Modelling

So far, morphological modelling has mainly focused on the analysis and hindcast of short-term and medium-term changes. In the last few decades, process-oriented modelling, i.e. based on descriptions of elementary wave, current and sediment transport processes, has developed very fast and is now reaching the level of fully-3D simulations. Modern observation and visualisation techniques help to increase the insight into what actually happens at the process level (e.g. Nelson et al. [25]).

Even if lack of computer power is not a problem, this approach has or may have some fundamental deficiencies when it comes to long-term prediction. Firstly, it is essentially deterministic, i.e. it operates in the space-time domain, but ignores the stochastic dimension that is inherent to the natural systems we are working with. The models are often driven with 'representative' conditions, such as the dominant discharge in the case of river models (e.g. Jansen [18]), a single or at most a few representative tides (e.g. Latteux [21]), or a limited set of representative wave conditions (e.g. De Vriend et al. [9]). At best, the models are driven with measured time series of input variables, thus taking into account the stochastic variation within such a series, but not the fact that these series are only one realisation of a stochastic process, the statistical properties of which are usually called 'climate' or 'regime'. In other words: the 'ensemble' dimension, which encompasses the other time series that could just as well have occurred within the same climate, is ignored. In this paper we will discuss some examples showing that this can be an essential shortcoming in long-term morphological predictions.

A second fundamental shortcoming of process-oriented modelling is associated with the non-linear character of the system. This leads not only to dynamic interactions between phenomena at different scales, but it also involves the possibility of multiple equilibrium states and deterministic chaos. Hence there may be inherent limits to the predictability of such systems, more or less similar to the case of turbulent flow. In a way, process-oriented morphological modelling is comparable to direct numerical simulation of turbulence, or, at a slightly higher level of aggregation, large-eddy simulation of turbulent flows: the model is able to predict the general structure and the statistical properties of the small-scale processes (mean velocities, Reynolds stresses, turbulence kinetic energy, length and time scales, coherent structures, etc.), but not their space-time evolution in a deterministic sense. An important difference between process-oriented morphological models and many turbulent flow simulations, however, is the occurrence of changes at higher scale levels. Only in the exceptional case of equilibrium at a certain scale level[3], there is statistical stationarity at that level. Therefore, time-averaging cannot always be used as a substitute to ensemble-averaging (defined as averaging over different realisations of a stochastic process at the same point in space and time).

The above predictability limitation can be inherent to the morphodynamic system (i.e. it occurs even under constant forcing), but it can also be provoked and/or aggravated by the random character of the forcing (cf. the randomly forced non-linear swing). Clearly, the amount of damping is an important aspect, which we have to include into the discussion on these matters.

A third point of attention when it comes to the predictive skill of a model concerns the uncertainty embedded in data, viz. (1) the data that are used as inputs into the model (forcing factors, parameters), or (2) the data that serve as a reference for model validation. Also, by introducing too many calibration parameters, a model can be over-fitted, i.e. it fits the reference data set very well, but has little other predictive skill than extrapolation under the assumption of *ceteris paribus* (the process just continues, without essentially changing). Also, introducing more uncertain data may well be counterproductive, as the increase of uncertainty in the model may exceed the increase of information. We will discuss some examples which are suspect in this respect.

We can conclude from the above, that morphological modelling has come a long way towards the ability to analyse and hindcast morphological phenomena, but that more work needs to be done in order to achieve long-term predictions that include quantitative indications of their predictive skill and the uncertainty involved.

8.6 Dealing with Uncertainty in the Forcing

8.6.1 Example: river constriction

In order to show what uncertainty in the external forcing can do and to what extent we are able to deal with it, we consider the case of a long[4] constriction of a straight river, which was originally in global equilibrium (Figure 8.4).

[3] In morphology, the concept of equilibrium cannot be applied in an absolute sense: it is scale-dependent. A tidal inlet system, for instance, can be in global equilibrium at a time scale of decades to centuries, whereas it undergoes significant changes, at smaller scales (e.g. the quasi-cyclic behaviour of channels; see Israel and Dunsbergen [17]), but also at larger scales (the gradual fill-up process at the century to millennia scale).

[4] *Long* means here: comparable to the length scale of the backwater curve, so several kilometres in the case of a lowland river.

This case was run 750 times with a 1-D model, each time with a different realisation of the discharge time series, drawn from a set with prescribed statistical properties that were derived from the Rhine regime. For a detailed description of this Monte Carlo exercise: see Van der Klis [42]).

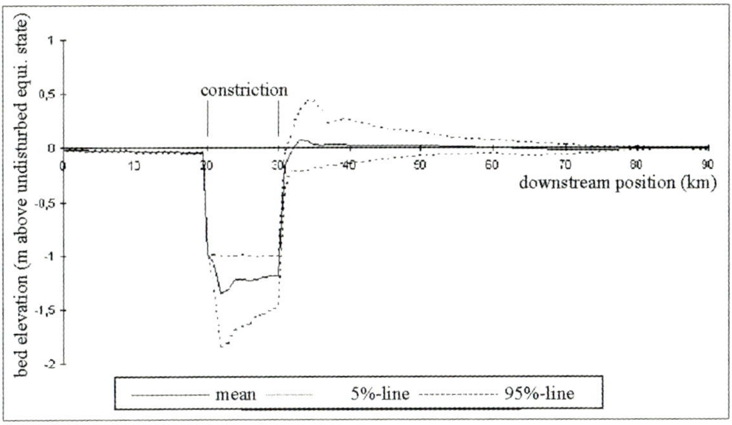

Fig. 8.4. Forcing-induced uncertainty in the case of a long constriction in a straight river [courtesy H. van der Klis].

Figure 8.4 shows the results of these computations, in terms of the ensemble-average and the upper and lower envelopes of the bed level in every point along the river. Note that neither of these lines necessarily represents an actual realisation of the bed topography. In the constriction area, the range of variation of the bed levels turns out to be of the same order of magnitude as the mean effect of the constriction. Moreover, a similar range is found in the first kilometres downstream of the constriction, even though the mean effect is virtually zero there. Clearly, this variation is caused by the deposition and the further transport of erosion products associated with erosion events in the constriction area, and by the pick-up of sediment and the subsequent migration of a scour wave which attend sedimentation events in that area.

The ensemble-averaged bed level turns out to be reproduced rather well with a dominant-discharge[5] model. The range, however, is fundamentally related to events and can therefore not be found from a constant-discharge model, not even from additional runs with extremely high and low discharges. At best can it be estimated by running a deterministic variable-discharge

[5] The dominant discharge is defined here as the constant discharge that yields the same yearly transport volume as found in nature.

In principle, the system's non-linearity can make it sensitive to the sequence of forcing events, since the response depends on the system's state[6]. If this chronology turns out to be practically important, indeed, the statistical information that is usually provided (pdf's, energy density spectra) is not sufficient to characterise the model input, since it contains no phase information. Bispectra provide a means to include this phase information, but they are not yet commonly used in morphological data representation. We should therefore assess the sensitivity of predictive model results to a range of statistical properties of the input conditions. Also should we address questions like how these properties can best be identified from measured data, and how one can generate input time series that represent these properties.

The fourth question refers not only to the effort needed to run complex multi-dimensional morphological models in a Monte Carlo setting, but also to the danger of relying exclusively on one source of information (in this case the model). Nature, however, usually provides only one realisation of the stochastic forcing process, i.e. it gives only one point on the ensemble-axis. Laboratory experiments, repeated with a variety of input scenarios, could be a viable alternative. This avenue deserves further exploration.

Apart from these research issues, there is another type of question to be addressed: once it is possible to predict statistical aspects of the morphology and to quantify uncertainty in those predictions, what information should those predictions provide to end-users? This probably requires a learning process at either end.

8.7 Dealing with Uncertainty in the Process

8.7.1 General

As indicated before, there is another aspect of predictive morphological modelling that originates from the non-linear character of morphodynamic systems. Non-linearity may give rise to inherent predictability limits, which are basically independent of the predictability of the forcing. Even under forcing conditions that are exactly known, the system may jump between different equilibrium states, or it may exhibit chaotic behaviour.

Weakly non-linear stability analyses have been carried out for various morphodynamic systems (e.g. Seminara and Tubino [36], for meandering rivers; Vittori and Blondeaux [45], for wave-induced bed ripples; Schielen et al. [34], for alternate bars in straight rivers; Vittori et al. [46], for beach cusps). Depending on the approach, these analyses yield a type of amplitude evolution equation that, in certain ranges of the parameter values, may involve quasi-periodicity and deterministic chaos (Doelman [10]). Recent fully non-linear

[6] This is not the case in linear systems, where the superimposition principle applies. Hence the response to any part of the forcing can be considered separately from the contributions from all other parts of the forcing and the initial and boundary conditions. The system's state is just the sum of all these contributions.

stability analyses of idealised models of the morphology of rivers, tidal inlets, estuaries and sandy coasts (e.g. Schuttelaars and De Swart [31,32], Calvete [5], Calvete et al. [6]) have shown that morphological systems can have multiple attractors, indeed. The solution may jump between these attractors, thus exhibiting a kind of periodicity. Chaotic behaviour of morphodynamic systems, however, has not been identified so far by this type of analyses.

8.7.2 Alternate bars in rivers: an example of quasi-uncertainty

An example of a weakly non-linear stability analysis that reveals non-trivial behaviour concerns alternate bars in a shallow straight river (Schielen et al. [34], Schielen [33], Knaapen et al. [19]). At a scale level of a few bar periods or lengths, these features look regular and seem to have a constant amplitude. At the level of many wavelengths or periods, their behaviour turns out to be quasi-periodic, i.e. the amplitude changes through time in a rather irregular way. Figure 8.6 shows an example of this quasi-periodic behaviour.

Fig. 8.6. Example of quasi-periodic behaviour of alternate bars in a straight alluvial channel [from: Schielen [33]].

Formally speaking, quasi-periodic behaviour is perfectly predictable. In practice, however, it would take very long model runs. Hence there is a tendency to treat quasi-periodicity as a stochastic phenomenon, characterised by its statistical properties. Clearly, a small number of deterministic model runs may help to determine these properties.

8.7.3 Example: channel/shoal formation in a coastal lagoon

Schuttelaars ([30]; also see Schuttelaars and De Swart [31]) analyses the hypothetical case of a short tidal embayment of rectangular planform. The longshore and cross-shore dimensions are assumed to be small as compared to the tidal wave length. A sinusoidal tidal water level variation is imposed at the seaward end. The water level is assumed horizontal (pumping-mode approach), but linearised friction is included in the computation of the velocity field, in such a way that the current model actually boils down to a friction-dominated rigid-lid model with an imposed time-variation of the lid level. The sediment transport is assumed to be purely diffusive (no advection), with the equilibrium concentration proportional to the velocity squared. Downhill gravitational transport is taken into account by adding a diffusive term to the sediment balance equation. The time scale of the morphological changes is assumed much larger than the tidal period.

According to this model, the system's basic state is a plane cross-shore sloping bed. This solution satisfies Le Conte's [22] - or O'Brien's [26] - linear relationship between the tidal prism and the cross-sectional area below mean sea level. Beyond a critical value of the friction factor, this basic state turns out to be linearly unstable, i.e. it is susceptible to exponentially growing harmonic perturbations. This instability must be attributed to a positive feedback between the bed topography and the sediment concentration, which tends to be higher in the deeper parts. The result is a tendency of a channel to form near the central cross-shore axis of the embayment, and for a shoal to develop along either closed side.

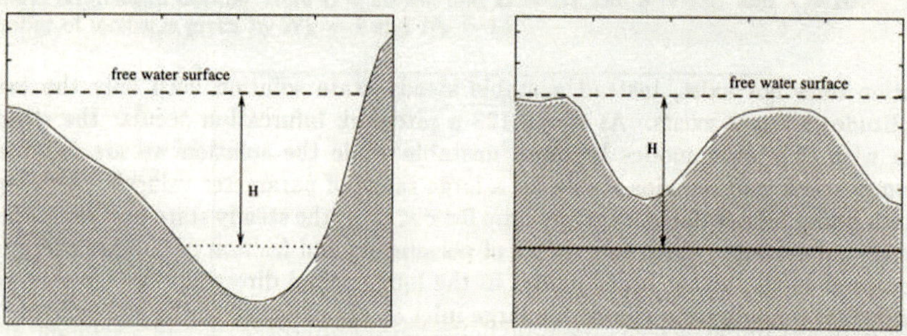

Fig. 8.7. Computed cross-sectional profiles of the bed in a rectangular embayment of 20 km long and 2.1 km wide, at 6 km from the mouth (left panel) and at 14 km from the mouth (right panel) [from: Schuttelaars [30]].

A fully non-linear stability analysis revealed that, beyond a certain (still rather small) value of the friction factor, the system has many different steady states, some of which are stable and others are not. Bifurcated channel patterns and periodic solutions (jumping from one steady state to another) are found, which remind of the natural quasi-cyclic migration of channel and shoal patterns. Figure 8.7 gives an example from a test computation.

Clearly, the model used here is highly simplified, if not over-simplified, and more research needs to be done (e.g. Van Leeuwen and De Swart [43]), but the non-linear analysis technique has proven to be a powerful tool in the analysis of this kind of morphodynamical systems.

8.7.4 Example: shoreface-connected ridges

Van de Meene [39,40] gives an extensive analysis the shoreface-connected ridges along the Dutch coast, but primarily from a hydrodynamical and sedimentological point of view. He showed that they are most probably morphologically active, instead of being geological relicts. Their top layer consists of recent storm deposits, but their core can be hundreds of years old. Important aspects are their orientation (under a sharp angle with the shore, opening towards the prevailing mean current direction).

Linear stability analyses of these features (e.g Falqués et al. [13]) have shown that they can be explained as instabilities which are trapped near the transition between the lower shoreface and the inner shelf. The shoreface slope is essential to their formation. Calvete ([5]; also see Calvete et al. [6]) performed a fully non-linear stability analysis of a storm-dominated shoreface and found that, depending on the coefficient of proportionality in the downhill gravitational transport component, there is no, one, or more than one initially unstable mode, and one or more attractor states. In contrast to what is usually assumed in linear stability analyses, the ultimate state is not always dominated by the initially fastest growing mode. There are also indications that quasi-periodic behaviour may occur.

The spatial patterns exhibit the typical non-linear features that are also known from ripple-patterns: downstream skewness, multiple scales and dislocations.

This illustrates that this type of patterns may evolve in different directions and jump quasi-periodically between different attractors. A deterministic process-oriented simulation model will have difficulty to reproduce this, the more so if the system is subject to a random forcing.

8.7.5 Offshore sand waves: another example?

New measuring techniques often reveal new phenomena. A few years ago, airborne radar techniques turned out to show features of the seabed topography (Alpers and Hennngs [2]; Vogelzang et al. [47]). The wavelength of these features was a few hundreds of metres, which suggested that these were sandwaves, but the complexity of the image formation process makes it impossible to tell what exactly the radar images reveal of the bed topography and what not. A preliminary comparison of SLAR-pictures and depth-soundings from the same area suggested a discrepancy in wavelength of about a factor two, i.e. the radar sees features of about twice the predominant wavelength in the soundings. Linear stability analyses like the ones described in the previous sections (Hulscher [16]; Van den Brink and Hulscher [41]) do explain the existence of wave-like instabilities with a wavelength of some hundreds of metres. Due to the uncertainty in the model parameters (e.g. sediment size, bed friction), however, one cannot claim that this model confirms or falsifies either of the two observations.

A recent application of a ship-borne multi-beam scanning technique, which gives much more detailed snapshot pictures of the bed topography, showed the co- existence of two modes, one about twice as long as the other. This cannot be explained without further analysis. Recent progress suggests that there be some improvement possible by including more or better described physical processes (cf. Blondeaux et al. [4], Komarova and Hulscher [20]), but non-linear stability analysis seems to be the best candidate to explain this phenomenon.

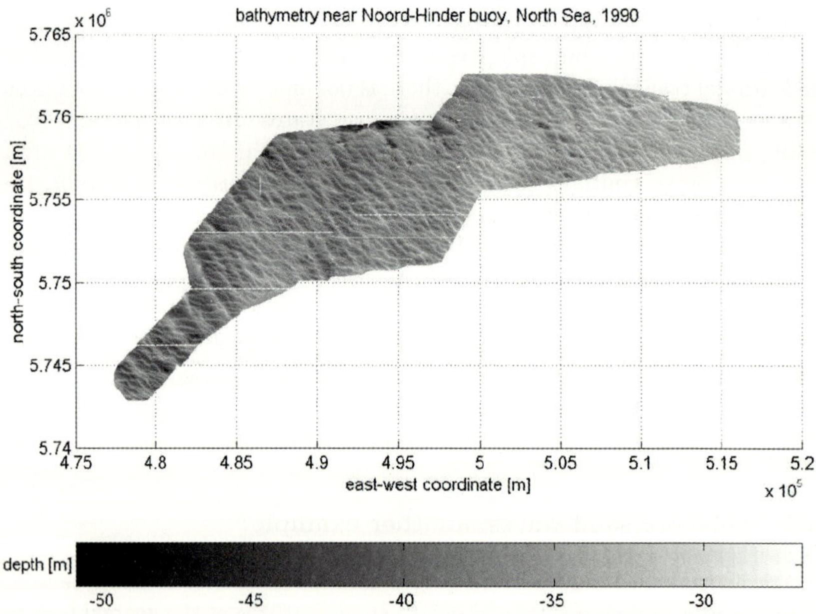

Fig. 8.8. Sandwave pattern in the access area to the Euro-channel, off Rotterdam, as revealed by multi-beam sonar [courtesy M.A.F. Knaapen].

8.7.6 Aggregation, the way to deal with predictability limits

If a system exhibits genuine predictability limits, e.g. because it becomes deterministically chaotic, this is not a reason to give up all hope of modelling its behaviour. Only can we no longer describe the system's state in deterministic terms, using a small-scale process-oriented model. We have to accept that the system's state can only be described in statistical terms.

This fact is known since long in turbulent flow modelling, where turbulence-averaged models are widely accepted. These models are formulated in terms of turbulence-averaged dependent variables, and include some form of empirical closure in order to deal with non-linear residual terms.

In general, formal averaging is one way of model aggregation. Other ways are:

- inclusion of empirical submodels, which lump complex small-scale processes into a simple relationship between larger-scale quantities. One example is the introduction of a sediment transport fromula, which lumps the complex grain motion in turbulent flow into a relationship between the net sediment flux and one or more mean flow parameters;
- ensemble-averaging, i.e. the model is (re)formulated in terms of statistical parameters (expectation, variance, etc.). One may claim that higher-order turbulence modelling belongs to this category. These models have not only the mean pressure and velocity components as dependent variables, but also other statistical properties (turbulence kinetic energy, Reynolds stress components, etc.). Even though these models describe statistical properties, they are essentially deterministic;
- stochastic modelling, in which the dependent variables and sometimes even the parameters are stochastic; in the case of morphological modelling, this is a largely unexplored area;
- projection onto a (usually orthogonal) function set, e.g. Fourier components, Chebyshev polynomials, EOF's, POP's. A well-kown example is wave modelling in the frequency-domain: the time-variation is covered by the basefunctions (progressive Fourier modes) and the model equations describe the spatial distribution - and sometimes the time-evolution at scales much larger than the wave period - of wavenumber, amplitude, direction, etc;
- model reduction via theoretical behaviour analysis, such as the stability analyses referred to in the previous sections. This ultimately result into a set of amplitude and phase evolution equations.

In various other fields of science (e.g. meteorology, oceanography), model aggregation is much more common than in morphological modelling. Most of the existing large-scale morphological model concepts have been developed more or less heuristically, or on the basis of empirical laws derived from measured data, without much of a connection with smaller-scale processes.

There are many indications of the existence of predictability limits in morphology (also see the previous sections), possibly even at a series of different scale-levels (scale-cascade; see De Vriend [7,8]). This is one good reason to claim that model aggregation is an important issue in morphological modelling. Another one is that we have to learn how to interpret morphodynamic behaviour at the appropriate scale level. There is not much point in referring to the Navier Stokes equations and a model of elementary grain-grain interactions if we have to interpret large-scale morphological changes in a given situation, let alone if we have to predict future developments after some interference.

8.7.7 Research strategy with respect to process-predictability

The cases discussed in the foregoing are just a few examples from a much wider field of research on the linear and non-linear behaviour of morphodynamic systems, with the universities of Genova and Utrecht and UPC Barcelona as prominent centres of activity. Such analyses, however simplified and disputable the underlying models may be, reveal much of the internal dynamics of these systems. Thus they have a lot to contribute to the body of morphological knowledge.

In fact, these theoretical behaviour analyses deserve a place in the morphological research methodology, at the same level as field measurements (monitoring programmes, as well as campaigns), laboratory experiments and numerical modelling. Metaphorically speaking: the table of morphological knowledge has four footings, instead of three!

A very strong and unique point of non-linear stability analyses is that they enable us to identify multiple solutions. One might attempt to use process-oriented numerical simulation models for this purpose (e.g. run with a variety of initial conditions, in order to trigger all possible solutions), but this remains searching for a needle in a haystack.

Knaapen et al. [19] try to avoid this problem by using a frequency-domain model, viz. a Ginzburg Landau model, as a link between the initial and the observed state of alternate bars in a straight channel (both defined in the physical space). Once the model has been fitted to the observed state, it is possible to reconstruct the corresponding initial state with a genetic algorithm. The resulting model, however, turns out to be extremely sensitive to inaccuracies in the observed state. More research is needed before this detour via the frequency-domain can be accepted or rejected as a practically viable option.

A combination of a theoretical behaviour analysis, such as linear or non-linear stability analyses, and a numerical model can be very strong, in that the numerical model can be utilised to test the effects of physical complications that have been omitted for simplicity from the analysis. Thus one can overcome the principal shortcomings of either approach: the (over-)simplification of the basic mathematical model in the analysis, and the impossibility to deal with multiple solutions in the numerical model. More research is needed, however, in order to make this operational.

Once this has been achieved, another step on the way to a practicable methodology for long-term morphological prediction can be made: the actual identification of predictability limits in a given situation. Before we start investing into aggregated models, we have to assess where and when they are needed.

The final step in this process is the development of aggregated model concepts, such that in any given situation we have access to a consistent set of models at different aggregation levels. Yet, this is not enough: along with the model development, we have to develop a good understanding of what happens at every relevant scale level. Being able to think in aggregated terms is just as important as having aggregated models available.

8.8 Dealing with Uncertainty in Parameters and Data

8.8.1 Example: bed roughness in a river

In most morphological models, the bed roughness is a user-supplied input parameter. The effective roughness, however, depends to a large extent on the small-scale bed morphology, which is part and parcel of the morphodynamic process. Hence, approximating the bed roughness as a prescribed constant is often an over-simplification of reality (also see Wilbers and Ten Brinke [51]).

This problem has been recognised quite some time ago by various modellers (see Engelund and Fredsøe [11]), who introduced dynamic roughness predictors to couple the bed roughness to hydrodynamic and sediment transport quantities. Such a roughness predictor is sometimes used as an explicit model component, sometimes it is embedded in the sediment transport model.

In order to illustrate the effect of including a roughness predictor into a morphological model, we consider the simple case of a straight alluvial channel of constant width, debouching into a deep lake with a constant water level. The river system is forced by a prescribed discharge, q, and sediment flux, s, two mutually independent quantities, both expressed in volume per unit with and time. Uniform flow is supposed to be described by Chezy's law,

$$u = C\sqrt{hi} \tag{8.1}$$

in which u is the flow velocity, C is Chezy's parameter, h is the water depth and i is the bottom slope. If the transport rate is given by the power-law formula

$$s = aC^{-3}u^5 \tag{8.2}$$

in which a is a constant, and if C, q and s are also constants, the equilibrium state follows from

$$h_{eq} = \left(\frac{s}{aC^{-3}}\right)^{-1/5} q \quad \text{and} \quad i_{eq} = \left(\frac{s}{aC^{-3}}\right)^{3/5} \frac{1}{C^2 q} \tag{8.3}$$

The elasticities of h_{eq} and i_{eq} with respect to q and s are given by

$$\frac{s}{h_{eq}}\frac{\partial h_{eq}}{\partial s} = -\frac{1}{5}\ ;\ \frac{q}{h_{eq}}\frac{\partial h_{eq}}{\partial q} = 1\ ;\ \frac{s}{i_{eq}}\frac{\partial i_{eq}}{\partial s} = \frac{3}{5}\ ;\ \frac{q}{i_{eq}}\frac{\partial i_{eq}}{\partial q} = -1 \tag{8.4}$$

and the sensitivity to the constant C is indicated by

$$\frac{C}{h_{eq}}\frac{\partial h_{eq}}{\partial C} = -\frac{3}{5}\ ;\ \frac{C}{i_{eq}}\frac{\partial i_{eq}}{\partial C} = -\frac{1}{5} \tag{8.5}$$

From this result one might be tempted to conclude that especially the equilibrium bed slope is not very sensitive to C, so that there is not much point in improving the description of this quantity.

If the following roughness predictor is introduced,

$$C = \beta h^n \tag{8.6}$$

both in the flow model and in the sediment transport formula, the equilibrium state follows from

$$h_{eq} = \left(\frac{s}{a\beta^{-3}}\right)^{-\frac{1}{5+3n}} q^{\frac{5}{5+3n}}\ ;\ i_{eq} = \left(\frac{s}{a\beta^{-3}}\right)^{\frac{3+2n}{5+3n}} \frac{1}{\beta^2}\frac{1}{q^{\frac{5+4n}{5+3n}}} \tag{8.7}$$

The elasticities of h_{eq} and i_{eq} with respect to q and s are now given by

$$\begin{aligned}\frac{s}{h_{eq}}\frac{\partial h_{eq}}{\partial s} &= -\frac{1}{5+3n}\ ;\ \frac{q}{h_{eq}}\frac{\partial h_{eq}}{\partial q} = \frac{5}{5+3n}\ ;\\ \frac{s}{i_{eq}}\frac{\partial i_{eq}}{\partial s} &= \frac{3}{5+3n}\ ;\ \frac{q}{i_{eq}}\frac{\partial i_{eq}}{\partial q} = -\frac{5+4n}{5+3n}\end{aligned} \tag{8.8}$$

In a similar way, one can evaluate the elasticities if the roughness predictor is included either in the flow model, or in the sediment transport formula only. Table 8.1 gives a summary of the results for $n = 0.5$, a rather extreme value, but nonetheless applied in practice (e.g. Enggrob and Thierry [12]). As an indication of what happens in the case of a variable discharge, Table 1 also gives the elasticities for a binary discharge distribution, 30% below average during 90% of the time, and five times as high during the remaining 10% of the time. Note that in this case the equilibrium depth only refers to the mouth of the river.

Table 8.1 reveals a substantial sensitivity, especially of the equilibrium bed slope (!), to the roughness predictor is included and the way it is implemented in the model. This effect is strongest for the binary discharge distribution, in which case even the sign of the elasticity factor $\delta(i_{eq}, \Delta q)$ may be reversed.

The most common approach to sensitivity analysis deals with each of the model parameters one by one, starting from a certain 'pivot' solution[7]. The

[7] This means that a basic solution with a certain parameter setting is chosen as a starting point, and that the parameters to which the sensitivity is to ve investigated are varied one by one, keeping all the others at their original value.

	O_1	FS_1	F_1	S_1	O_2	FS_2	F_2	S_2
$\delta(h_e, s)$	-0.2	-0.15	-0.2	-0.15	-0.2	-0.15	-0.2	-0.15
$\delta(h_e, q_m)$	1	0.77	1	0.77	0.28	0.22	0.28	0.22
$\delta(h_e, \Delta q)$	-	-	-	-	0.72	0.55	0.72	0.55
$\delta(i_e, s)$	0.6	0.62	0.8	0.46	0.6	0.62	0.8	0.46
$\delta(i_e, q_m)$	-1	-1.08	-2	-0.31	-0.51	-0.52	-0.87	-0.25
$\delta(i_e, \Delta q)$	-	-	-	-	-0.29	-0.35	-1.13	0.04

Table 8.1. Elasticities of the equilibrium depth and the equilibrium slope with respect to the water and sediment fluxes, in the case of a straight channel debouching into a deep lake with constant water level (left part); right part: elasticities with respect to the sediment flux, the mean discharge and the discharge difference in the same case with a combination of two discharges (base flow during 90% of the time, discharge 5 times as high during 10% of the time). Legend: δ (a,b) = elasticity of a with respect to b; O = no roughness predictor; FS = roughness predictor in flow and transport model; F = roughenss predictor in flow model only; S = roughness predictor in transport model only.

above example illustrates that this does not always yield a correct indication of the marginal profit (in terms of uncertainty reduction) of replacing such a parameter by a refined submodel.

The table also shows that the morphology is rather sensitive to the bed roughness predictor and the way it is embedded in the model. Enggrob and Thierry [12] arrive at a similar conclusion from their simulation of a river braiding process. As a consequence, the system is also sensitive to uncertainties in the bed roughness submodel. Given the state-of-the-art in roughness prediction, these uncertainties are considerable. Moreover, we don't even know to what extent they are inherent to the system, or attributable to our lack of knowledge. How their effects compare to those caused by the uncertainty in the discharge remains to be investigated.

8.8.2 Example: bed composition in a river

In most morphological models, the sediment is implicitly assumed to be uniform, which is seldomly in line with reality. Especially in situations with strong gradients in the velocity and transport fields, there is usually sediment sorting by grain size and/or density. This sorting takes place via complex exchange processes in the water column (in the case of suspended load transport) and in the top layer of the bed. For recent developments at this point, see Blom and Ribberink [3] and Parker [28]).

Models of this kind of sediment sorting have been proposed some time ago (e.g. Ribberink [29], Holly and Rauel [14]), but they are still not robust under all conditions, probably due to an insufficient description of the underlying physics. An alternative that is sometimes used (cf. Mosselman et al. [24]) is to prescribe the spatial distribution of the grain size at the bed surface,

to predictive morphological modelling, not only for its direct use (better controlled predictions), but also because of the feedback effects: once these skill-measures are available and have been accepted, research will focus more directly on increasing the predictive skill.

8.9 Conclusion

We can conclude from the foregoing that long-term predictive morphological modelling is a research and development field of its own, beyond the traditional world of model fitting, hindcasting and deterministic 'extrapolation' into the future. Learning how to deal with uncertainty and make it visible, instead of ignoring it, is vital to the further development of this field.

Important sources of uncertainty are the forcing conditions (in which weather and climate usually play a role), the model concept (all necessary processes included? at the right aggregation level?), the non-linear dynamics, the model parameters and the calibration and verification data. We need to develop techniques to evaluate how these sources influence the uncertainty in the predictions, i.e. the interpreted model results.

More concretely, important research issues in this domain are:

- developing experience, insight and analysis techniques concerning the effects of uncertainty in the forcing; increasing the awareness of the ensemble-dimension and learning how to deal with it; investigating the possibility to utilise laborararatory experiments to explore this dimension in complex real-life situations;
- pursuing the further development and application of non-linear behaviour analyses of idealised morphological systems, in combination with numerical model studies and field data analyses, to underpin the simplifying assumptions and to put the conclusions into a real-life perspective; explicit assessment of the existence or non-existence of inherent predictability limits at practically relevant scale levels in a number of cases, representative of practical problems involving long-term morphological changes;
- for various classes of long-term pediction problems: development of a consistent cascade of model concepts at different levels of aggregation, in order to overcome inherent, external and also practical predictability limits;
- developing methods for multi-parameter sensitivity analysis of non-linear models with large uncertainty ranges (low signal-to-noise ratio);
- developing quantifiable measures of predictive skill, as well as methods to analyse models at this point, given the various sources of uncertainty and the tuning parameters.

Resuming the foregoing metaphore: the table of morphological knowledge has four footings, but also a blade: uncertainty analysis.

Clearly, the above does not mean that process-research should be abandoned, or even reduced. As long as new model applications keep on revealing the need for inclusion of new small-scale process descriptions, process research has a key role to play. This paper hopefully makes clear, however, that small-scale process-research alone is not enough to make progress in long-term predictive modelling.

8.10 Acknowledgements

Many of the ideas underlying this paper have been developed in a series of projects in the framework of the EU-MAST programme, especially G6-M (contract MAST 0035-C), G8-M (contract MS2-CT92-0027) and PACE (contract MAS3-CT95- 0002), and in the project "Intermediate Scale Coastal Behaviour: Measurement, Modelling and Prediction" funded under the NICOP-programme of the US Office of Naval Research. Further development is now taking place in the discussions around the Delft Cluster research programme, Theme "Coast and River".

The author wishes to thank the organisers of the RCEM-symposium for creating a most stimulating environment to discuss and further develop these ideas. He gratefully acknowledges the willingness of his Ph.D. students Hanneke van der Klis and Michiel Knaapen to make some of their recent results available for inclusion in this paper.

References

1. Aarninkhof, S.G.J., Hinton, C., Wijnberg, K.M. (1998) On the predictability of nearshore bar behaviour. In: B.L. Edge (Ed.): Coastal Engineering 1998. ASCE, New York, 2409-2422.
2. Alpers, W., Hennings, I. (1984) A theory of the imaging of underwater bottom topography by real and synthetic aperture rada. J. Geophys. Res. C89, 10529-10546.
3. Blom, A., Ribberink, J.S. (1999) Non-uniform sediment in rivers: vertical sediment exchange between bed layers. In: Proc. IAHR-Symp. on River, Coastal and Estuarine Morphodynamics, Univ. Genova, Dept. Env. Eng., Genova, 45-54.
4. Blondeaux, P., Brocchini. M., Drago, M., Iovenitti, L., Vittori, G. (1999) Sand wave formation: preliminary comparison between theoretical prediction and field data. In: Proc. IAHR-Symp. on River, Coastal and Estuarine Morphodynamics, Univ. Genova, Dept. Env. Eng., Genova, 197-206.
5. Calvete, D. (1999) Morphological stability models: shoreface-connected ridges. Ph.D. thesis, Universitat Politcnica de Catalunya, Barcelona, Spain, 159 pp.
6. Calvete, D., Falqués, A., de Swart. H.E., Dodd, N. (1999) Nonlinear modelling of shoreface-connected sand ridges. In: N.C. Kraus and W.G. McDougal (Eds.): Coastal Sediments '99, ASCE, New York, 1123-1138.

7. De Vriend, H.J. (1997) River Morphology: a manifestation of nonlinear dynamics. In: F.M. Holly, A. Alsaffar, M. English and A. Szollosi-Nagy (Eds.): Managing Water: Coping with Scarcity and Abundance, Proc. XXVIIth IAHR Congress - Theme A, ASCE, New York, 10-15.
8. De Vriend, H.J. (1999) Long-term morphodynamics of alluvial rivers and coasts. In: G. Bianchi (ed.): Environmental Applications of Mechanics and Computer Science, CISM Courses and Lectures no. 409, Springer Wien New York, 1-19.
9. De Vriend, H.J., Capobianco, M., Chesher, T., De Swart, H.E., Latteux, B., Stive, M.J.F. (1993) Approaches to long-term modelling of coastal morphology: a review. Coastal Eng., **21** (1-3), 225-269.
10. Doelman, A. (1990) On the non-linear evolution of patterns. Ph.D., Utrecht University, The Netherlands, 120 pp.
11. Engelund, F., Fredsøe, J. (1982) Sediment ripples and dunes. Ann. Rev. Fluid Mech. **14**, 13-37.
12. Enggrob, H.G., Tjerry, S. (1999) Simulation of morphological characteristics of a braided river. In: Proc. IAHR-Symp. on River, Coastal and Estuarine Morphodynamics, Univ. Genova, Dept. Env. Eng., Genova, 585-594.
13. Falqués, A., Calvete, D., Montoto, A. (1998) Bed-flow instabilities of coastal currents. In: J. Dronkers and M.B.A.M. Scheffers (Eds.): Physics of Estuaries and Coastal Seas, Balkema, Rotterdam, 417-424.
14. Holly, F.M. Jr., Rahuel, J.-L. (1990) New numerical/physical framework for mobile-bed modelling. Part 1: Numerical and physical principles. J. Hydr. Res. **28** (4), 401-416.
15. Holman, R.A., 1999. Pattern formation in the nearshore. In: Proc. IAHR-Symp. on River, Coastal and Estuarine Morphodynamics, Un. Genova, Dept. Env. Eng., Genova, p. 403-404 (full paper in this volume).
16. Hulscher, S.J.M.H. (1996) Tidally induced large-scale regular bed form patterns in a three-dimensional shallow water model. J. Geoph. Res. **101** (C9), 20,727-20744.
17. Israel, C.G., Dunsbergen, D.W. (1999) Cyclic morphological development of the Ameland Inlet, The Netherlands. In: Proc. IAHR-Symp. on River, Coastal and Estuarine Morphodynamics, Univ. Genova, Dept. Env. Eng., Genova, 705-714.
18. Jansen, P.Ph. (Ed.) (1979). Principles of River Engineering. Pitman Publ. Ltd., London, 509 pp.
19. Knaapen, M.A.F., Hulscher, S.J.M.H., De Vriend, H.J. (1999) On the possibility of genetic algorithms and the Ginzburg-Landau equation to model alternate bars: a sensitivity analysis. In: Proc. IAHR-Symp. on River, Coastal and Estuarine Morphodynamics, Univ. Genova, Dept. Env. Eng., Genova, 283-292.
20. Komarova, N.L., Hulscher, S.J.M.H. (2000) Linear instability mechanisms for sand wave formation. Accepted for publication in J. Fluid Mech.
21. Latteux, B. (1995) Techniques for long-term morphological simulation under tidal current action. Marine Geology, **126**, 129-141.
22. LeConte, L.J. (1905) Discussion on River and Harbour Outlets, "Notes on the improvement of river and harbour outlets in the United States", paper no. 1009 by D.A. Watts, Trans. ASCE, **55**, 306-308.
23. Migniot, C. (1997) Restoration of the maritime character of the Mont-Saint-Michel: general synthesis of existing knowledge on the hydro-morphological problems. Dir. Dépt. de l'Equipement de la Manche, Report Mission Mont Saint Michel, 89 pp (in French).

24. Mosselman, E., Sieben, A., Sloff. K., Wolters, A. 1999 Effect of spatial grain size variations on two-dimensional river bed morphology. In: Proc. IAHR-Symp. on River, Coastal and Estuarine Morphodynamics, Univ. Genova, Dept. Env. Eng., Genova, 499-507.
25. Nelson, J.M., Schmeeckle, M., Shreve, R.L., McLean, S. (1999) Sediment entrainment and transport in complex flows. In: Proc. IAHR-Symp. on River, Coastal and Estuarine Morphodynamics, Univ. Genova, Dept. Env. Eng., Genova, 3-4 (full paper in this volume).
26. O'Brien, M.P. (1931) Estuary tidal prism related to entrance areas. Civ. Eng. **1** (8), 738-739.
27. Oost, A.P. (1995) Dynamics and sedimentary development of the Dutch Wadden Sea, with emphasis on the Frisian Inlet. Univ. of Utrecht, Geologica Ultraiectina 126, 454 pp.
28. Parker, G. (1999) Meandering channels in the submarine environment. In: Proc. IAHR-Symp. on River, Coastal and Estuarine Morphodynamics, Univ. Genova, Dept. Env. Eng., Genova, 531-532 (full paper in this volume).
29. Ribberink, J.S. (1987) Mathematical modelling of one-dimensional morphological changes in rivers with non-uniform sediment. Ph.D. thesis, Delft Univ. Techn., Delft, The Netherlands, 200 pp.
30. Schuttelaars, H.M. (1997) Evolution and stability analysis of bottom patterns in tidal embayments. Ph.D. thesis, Utrecht University, The Netherlands, 134 pp.
31. Schuttelaars, H.M., De Swart, H.E. (1999a) Formation of channels and shoals in a short tidal embayment. J. Fluid Mech. **386**, 15-42.
32. Schuttelaars, H.M., De Swart. H.E. (1999b) Multiple morphodynamic equilibria in a one-dimensional tidal embayment. In: Proc. IAHR-Symp. on River, Coastal and Estuarine Morphodynamics, Univ. Genova, Dept. Env. Eng., Genova, 619-628.
33. Schielen, R.M.J. (1995) Non-linear stability analysis and pattern formation in morphological models. Ph.D. thesis, Utrecht University, The Netherlands, 150 pp.
34. Schielen, R., Doelman, A., De Swart, H.E. (1992) On the nonlinear dynamics of free bars in straight channels. J. Fluid Mech. **252**, 325-356.
35. Seminara, G. (ed.) (1999) Proc. IAHR-Symp. on River, Coastal and Estuarine Morphology, Univ. of Genova, Dept. Env. Eng., Genova, 734 + 386 pp. (2 volumes).
36. Seminara, G., Tubino, M. (1989) Alternate bars and meandering; free, forced and mixed interactions. In: Ikdea, S. and Parker, G. (Eds.): River Meandering, AGU Water Res. Monograph 12, 267-320.
37. Simpson, D. (1998) Linking nearshore depth and shoreline change around Grays Harbour. In: G. Gelfenbaum and G.M. Kaminsky: Southwest Washington Coastal Erosion Workshop Reort 1998, USGS Open-File Report 99-524, 77-85.
38. Southgate, H.N., Capobianco, M. (1997) The role of chronology in long term mor-phodynamics: theory, practice and evidence. In: E.B. Thornton (ed.): Coastal Dynamcis '97, ASCE, New York, 943-952.
39. Van de Meene, J.W.H. (1991) Shoreface connected ridges along the Dutch coast. In: N.C. Kraus, K.J. Gingerich and D.L. Kriebel (Eds): Coastal Sediments '91, ASCE, New York, 512-526.

40. Van de Meene, J.W.H. (1994) The shoreface-connected ridges along the central Dutch coast. Ph.D. Thesis. KNAG/Netherlands Geographical Studies 174, Utrecht, The Netherlands, 222 pp.
41. Van den Brink, G.M., Hulscher, S.J.M.H. (1999) Prediction of sand waves and sand banks in the North Sea. In: Proc. IAHR-Symp. on River, Coastal and Estuarine Morphodynamics, Univ. Genova, Dept. Env. Eng., Genova, 217-226.
42. Van der Klis, H. (2000) Stochastic modelling of river morphology: a case study. Paper to be presented at the 8th Int. Symp. on Stochastic Hydraulics, Bejing, China, July 2000.
43. Van Leeuwen, S., De Swart, H.E. (1999) The effect of non-linear and deposition processes on the morphodynamics of tidal embayments. In: Proc. IAHR-Symp. on River, Coastal and Estuarine Morphodynamics, Univ. Genova, Dept. Env. Eng., Genova, 639-648.
44. Visser, P.J., Havinga, H., ten Brinke, W.B.M. (1999). How do we keep the river navigable? Land en Water **9** 25-27 (in Dutch).
45. Vittori, G., Blondeaux, P. (1992) Sand ripples under sea waves, Part 3: brick pattern ripple formation. J. Fluid Mech. **239**, 23-45.
46. Vittori, G., De Swart, H.E., Blondeaux, P. (1999) Crescentic forms in the nearshore region. J. Fluid Mech. **381**, 271-303.
47. Vogelzang, J., Wensink, G.J., Van Halsema, D., Van der Burg, G. (1990) Sea bottom topography with imaging radar. In: B.L. Edge (Ed.): Coastal Engineering 1990 Proc., ASCE, New York, 2279-2290.
48. Vrijling, J.K., Meijer, G.J. (1992) Probabilistic coastline position computations. Coastal Engineering, **17** (1/2), 1-23.
49. Vroon, J.H., Storm, C., Coosen, J. (1997) Western Scheldt, stiff or sturdy? Rijkswaterstaat - National Institute of Coastal and Marine Management, OOST-WEST evaluation report.
50. Wijnberg, K.M., Terwindt, J.H.J. (1995) Extracting decadal morphological behaviour from high- resolution, long-term bathymetric surveys along the Holland coast using eigenfunction analysis. Marine Geol. **126**, 301-330.
51. Wilbers, A.W.E., Ten Brinke, W.B.M. (1999) In: Proc. IAHR-Symp. on River, Coastal and Estuarine Morphodynamics, Univ. Genova, Dept. Env. Eng., Genova, 303-312.

9 River and Tidal Networks

Andrea Rinaldo, Stefano Lanzoni, Marco Marani

Dipartimento di Ingegneria Idraulica, Marittima e Geotecnica, Università di Padova, via Loredan 20, I-35131 Padova, Italy

9.1 Abstract

The modern analysis of channel patterns in nature over a wide range of scales, be it a tree-like river, braided landforms or the complex incisures of a tidal network, relies on accurate digital mapping technologies and on objective manipulations of remotely sensed imaging. Such analyses reveal extraordinary diversity of natural forms, and yet deep regularity and symmetry, regardless of a variety of environmental factors - what strikes is diversity of hydro- and morpho-dynamics, adapting to (and coevolving with) geology, climate, vegetation or exposed lithology. Our observations strongly suggest that river networks are indeed a paradigm of scale-invariant, or fractal, forms ubiquitous in nature, whereas tidal networks bear rather strikingly the signatures of scale-dependence. Here we review a somewhat subjective choice of linked results, chiefly observational, that are deemed helpful towards a theory of the dynamic origin of scale invariance in the fluvial landscape referring to a common mechanism of growth and stabilization of open, dissipative systems with many degrees of freedom.

9.2 Introduction and Background

A foremost question pertaining to science at large concerns the dynamic origin of scale invariance in natural patterns and forms. In this general framework, the fields of geomorphology and hydrology play an important role because the 'network' concept is directly relevant to the topic of scale invariance. In fact, two related problems have interested scientists for a long time. One is the fundamental dynamic reason behind Mandelbrot's observation that many structures in nature - such as river networks or coastlines - are fractal, i.e. looking 'alike' on many length scales. The other is the origin of the widespread phenomenon called $1/f$ noise, originally referring to the particular property of a time signal, be it the light curve of a quasar or the record of river flows, which has components of all durations, i.e. without a characteristic time scale. The name $1/f$ refers to the power law decay with exponent (-1) of the power spectrum $S(f)$ of certain self-affine records and it is conventionally extended to all signals whose spectrum decays algebraically.

Power-law decay of spectral features is also viewed as a fingerprint of spatially scale-free behaviour, commonly defined as critical. In this framework criticality of a system postulates the capability of communicating information throughout its entire structure, connections being distributed on all scales. It has been suggested that the abundance in nature of spatial and temporal scale-free behaviours may reflect a universal tendency of large, driven dynamical systems with many degrees of freedom to evolve into a stable critical state, far from equilibrium, characterized by the absence of characteristic spatial or temporal time scales. This key idea and its successive applications (e.g. Bak [1]) address such universal tendency towards a scale-free state and bear important implications on our understanding of complex natural processes.

The common dynamic denominator underlying fractal growth is now central to our interests in landform evolution. The fluvial context is perhaps significant to the above, owing to the quality and range of data and the solid (and verified) theoretical background. In fact, in many branches of science, and in geomorphology in particular, a reductionist approach is often pursued to understand the physics of the processes. The reductionist tenet is that if one is capable to dissect and understand the processes to the smallest scales characteristics of the investigated phenomenon then the capability to explain the general picture, including complexity, is granted. However, the reductionist approach, affected as it is by the need of specifying so many detailed processes operating in nature and the tuning of many parameters, though suited to describe individual forms, can hardly explain the ubiquity of scaling forms and the recursive characters of processes operating in very different conditions. The approach based on self-organized criticality, on the other hand, overcomes this limitation by focusing on general properties of the dynamics of large interactive systems rather than on the unlikely fine-tuning of its elementary ingredients to see whether self-organization emerges.

One crucial feature of the organization of fractal structures in large dynamical systems is the power-law structure of the probability distributions characterizing their geometrical properties. This behaviour, characterized by events and forms of all sizes, is consistent with the fact that many complex systems in nature evolve in an intermittent, burst-like way rather than in a smooth, gradual manner and/or are scale-free in space.

The hydraulic networks we analyze need some preliminary definition. In this context it would be more appropriate to refer to hydraulic trees (i.e. where a unique pattern leads from any site to a seed, or outlet regardless of the definition of the lattice of sites connected by the network structure). The runoff-generating region of fluvial basins (Schumm [32]; Rodriguez-Iturbe and Rinaldo [28]) produces tree-like drainage structures in a wide range of scales (Figure 9.1), from the lower cutoff dictated by prevailing hillslope processes to the upper cutoff where mesoscale fluvial processes or geologic constraints operate. At scales smaller than that defining drainage density, in fact, multiple-flow directions reflect topographic convexity whereas in topographically convergent areas aggregated flow results from advective geomorphic processes -

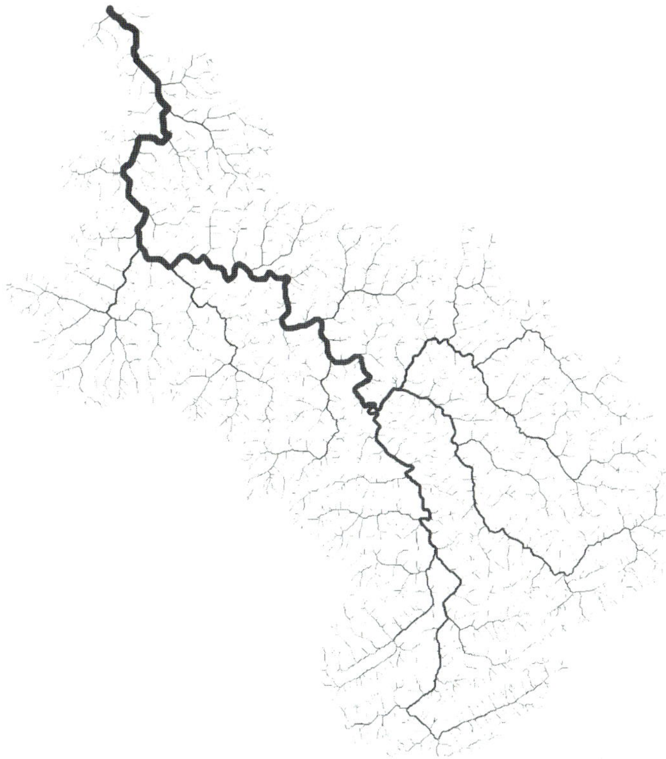

Fig. 9.1. One example of natural river network extracted from high accuracy digital terrain maps (Rigon, 2000, personal communication). The general idea is that without a scale bar it is impossible to devise even the approximate scale of a planar river form or of a topographic map, an ingredient essential to any form of scale-invariance. Although this is true for a wide range of scales, there exist lower and upper limits to landscape dissection into distinct valleys. The lower cutoff is generated by a threshold of channelization that sets a finite scale to the landscape. The upper cutoff is usually of geologic nature. Nevertheless it is remarkable that scale invariance holds from the order of 10 metres to the tens of thousands of kilometres (e.g. Rodriguez-Iturbe and Rinaldo [28]).

and trees emerge over a wide range of scales, typically from $O(10^2)$ to $O(10^6)$ m.

Tidal networks, i.e. carved in tidal environments by periodically oscillating flows and a variety of transport phenomena where biological processes play a major role, are also mostly tree-like structures (e.g. Fagherazzi et al. [4]), although shaped by processes radically different from the fluvial counterparts Figure (9.2). On the contrary, the tree-like framework clearly excludes braided

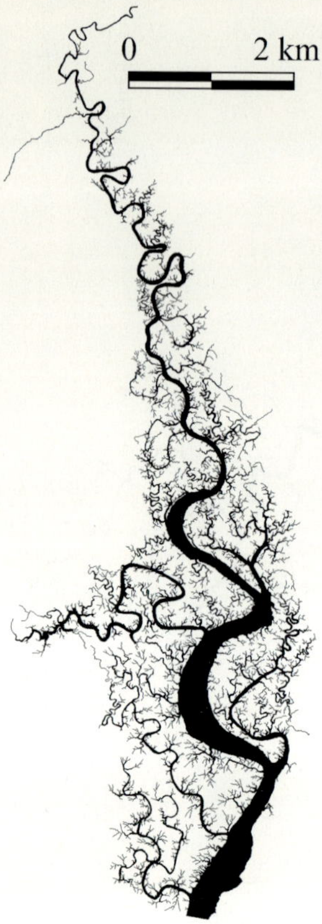

Fig. 9.2. One example of tidal network (after Rinaldo et al. [26]). Planar reconstruction of the Petaluma march from historical maps. Here the digital set of channelized pixels is obtained with resolution of 10×10 m^2.

patterns, per se the object of intense screening recently especially in relation to scaling theories (e.g. Murray and Paola [18]; Sapozhnikov and Foufoula Georgiou [30]).

In this paper we critically compare some recent results on the observational scale invariance of river networks across several orders of magnitude (e.g. Rodriguez-Iturbe and Rinaldo [28]) with observations from tidal environments (Fagherazzi et al. [4]; Rinaldo et al. [26]). A factor we keep in mind is the possible emergence of a coherent theoretical framework for the dynamic origin of fractals. Theory may thus provide a broad foundation to the linkages of scale-invariant forms, so common both in the living (e.g. McMahon and Bonner [15]) and the inanimate world (Mandelbrot [13]; Niemeyer et al. [19]; Sander [29]), with general features of the embedded evolutionary dynamics. The accessibility of large data sets, objectively collected, of acceptable precision and spanning the natural phenomena over a wide range of scales plays also an important role (Rodriguez-Iturbe and Rinaldo [28]; Rinaldo et al. [25]). Such a role, owing in particular to digital mapping techniques, provided hydrologists and geomorphologists with a unique opportunity for the analysis and the testing of theories and methods. As we now fully realize, many geomorphological relationships empirically known - some dating from the last century - carry the signatures of fractal growth and of critical self-organization (Mandelbrot [13]; Bak [1]; Rinalto et al. [24]) and, interestingly, of climatic changes (Rinaldo et al. [23]).

9.3 River Networks

We will not deal here with the appropriate techniques for extracting a fluvial network from a general digital terrain map, e.g. suitably differentiating areas characterized by topographic convergence, from (say, channelized colluvium) areas where topographic curvature is convergent. The reader is referred to Montgomery and Dietrich [16,17], Dietrich et al. [2], Howard [7] and to Rodriguez-Iturbe and Rinaldo [28] for a synthesis of fluvial patterns. We shall therefore assume that the procedure is reliable over a significant range of scales and provides accurately described features of natural networks (Figure 9.1).

Hydraulic networks are paradigms of scale-invariant geometrical forms, i.e. statistically similar structures where the part replicates the whole (almost) endlessly. One crucial feature of the organization of scale-invariant structures is the power-law structure of the probability distributions characterizing their geometrical properties (Mandelbrot [12,13]). For spatial structures like river networks a power-law probability distribution fits many measurable geometric quantities. The embedded scale invariance can be described through linked scaling exponents which, following Maritain et al. [14] and Rigon et al. [22], will be assumed as distinctive tools for the comparison of different channel networks. In fact, a far from superficial question that one addresses when comparing channel networks is what uniquely characterizes them. Certainly Hortonian or Tokunaga side branching statistics do not (Kirchner [8]; Rinaldo et al. [25]).

In addressing the common properties of a tree, many of the properties of interest are surrogated by the organization of total contributing areas. In a river network the total contributing area of a site i, i.e. a_i, is identified through drainage directions (topographic gradients) and is measured by the number of sites upstream of the site connected by the network. If N are the sites, the equation for total contributing area at the i-th site, a_i, is

$$a_i = \sum_{j \in N} a_j \, w_{ij} + r_i \tag{9.1}$$

where: N is the set of nodes of the network; r_i is the injection term, usually taken as unity (i.e. implying a constant elementary or pixel area); w_{ij} is the element of the adjacency matrix W. In the case of trees w_{ij} has only integer values, i.e. it equals 1 if i collects flow from its neighbour j through a drainage direction (i.e. if $j \to i$), and 0 otherwise. In the generic case of multiple flow directions w_{ij} is a real number satisfying the continuity constraint $\sum_j w_{ij} = r_i$. The $N \times N$ matrix W is defined by the elements $\{w_{ij}\}$, and one may define important properties from its eigenvalues. For example, $W^L = 0$ implies that $L-1$ steps define the maximum distance from source to outlet within an arbitrary complex tree. In matrix notation one has $a = (I - W^T)^{-1} r$ and the eigenvalues solve $det(\lambda I - W) = 0$ (where I is the identity matrix). In the network case one can show that all eigenvalues λ are zero.

The point value r_i added in equation (9.1) specifies the open, nonlocal character of the equation and defines the external mass injection at the i-th pixel (mass is simply area in the geometric description; it mimics flow rates under reasonable assumptions, see sect. 9.4). Thus at any site of a network one obtains a single value of a. In real networks (Rodriguez-Iturbe et al. [27]; Rodriguez-Iturbe and Rinaldo [28]) a is a random variable described by the finite-size scaling probability distribution:

$$p(a) \propto a^{-(1+\beta)} \mathcal{F} \left(a / L_{\parallel}^{1+H} \right) \tag{9.2}$$

which implies probability of exceedence of the form $P[\geq a] \propto a^{-\beta}$ (Rodriguez-Iturbe et al. [27]; Maritan et al. [14]); $L_{\parallel}$ is a linear size characteristic of the network, technically defined as the longest distance from a site on the boundary to the outlet; $\mathcal{F}$ is a scaling function whose detailed form is irrelevant but whose limits allow important comparisons; H is a (Hurst) exponent characterizing the elongation of the total area occupied by the network. The coefficient β characterizes the entire aggregation pattern. The observational result of Rodriguez-Iturbe et al. [27] is $\beta = 0.43 \pm 0.02$, which is an important quantity relatively independent of size, vegetation, geology, climate or orientation of the basin. Notice that the value of the slope β is unaffected by the size of the support threshold used in the identification of the network, of course within limits as care should be exerted in actually dealing with channelized pixels rather than with the contributing hillslopes. The coefficient H in real basins is in the range $0.75 - 0.80$ (Rodriguez-Iturbe and Rinaldo [28]). From the previous finite-size argument Maritan et al. [14] obtained the exact result $< a^n > \propto L_{\parallel}^{(n-2-\beta)(1+H)}$ (where $< a^n >$ is the n-th moment of the distribution of the random variable a) which allows the introduction of the relationship

$$\frac{< a^n >}{< a^{n-1} >} \propto L_{\parallel}^{1+H} \tag{9.3}$$

and hence a direct (and indeed quite robust) calculation of H from data.

We also measure distinctive properties of the length of the channels. It is appealing to define stream length properties without linking the analysis to any particular ordering scheme. One way to do this is to analyze the random variable defined as the longest distance measured through the network from a randomly chosen point to the boundary of the basin, i.e. the mainstream length L extended to the subbasins rooted in all sites of the network. Technically one defines the mainstream pattern upstream of any junction following the site having maximum area (in case of equal contributions one chooses at random) until a source is reached. The finite-size distribution of mainstream lengths L at any point in the basin for a given area a has scaling exponent ξ (Rigon et al. [22]) and obeys the same requirements of that of areas. The distribution is shown to be in the form $P[\geq l] = l^{-\xi} \mathcal{G}(l/a^h)$, where P is the probability distribution exceeding the arbitrary length l given a drainage area a, $\mathcal{G}$ is a scaling function whose specialization is immaterial. The ξ scaling exponent is in the observational range 0.75 - 0.90 (Maritan et al. [14]; Rigon et al. [22]). Notice that consistency requires that $\xi = \beta/h$. Notice also, in analogy with the previous result, that one has $< l^n > \propto a^{h(n-\xi-2)}$, i.e. the n-th moment of the distribution of mainstream lengths for a given area A scales with area A (Rigon et al. [22]), and

$$\frac{< l^n >}{< l^{n-1} >} \propto A^h \tag{9.4}$$

(see, for details, Rodriguez-Iturbe and Rinaldo [28]) pp.182-195). Experimental evidence strongly supports the above assumptions. Figure 9.3 shows the experimental verification (after Rigon et al. [22]) of eq. (9.4) for two real basins.

In conclusion, a set of independently-measured and theoretically linked scaling exponents distinctively characterize a river network. They define the topological structure through the properties of its aggregation, the length of the individual components not seen by topology alone, the degree of irregularity of the mainstream and the general elongation of the drainage area. Thus we will assume that a synthetic shape matches natural features when *all* scaling exponents fall in the suitable range (Table 9.1), not just one or two.

One notes (Rinaldo et al. [25]) that chance-dominated, random walk-type of models, once thought to capture the essence of Nature's mechanisms in producing networks, do not really reproduce observational features besides some superficial similarities. A complete and revealing comparison of the scaling coefficients for Leopold/Eden (Leopold and Langbein [10]; Leopold et al. [11]; Eden [3]; Howard [6]) and Scheidegger [31] networks with observational data is shown in Table 9.1. Shreve's [33,34] topological random networks also do not reproduce well the scaling exponents, although they reproduce less stringent statistics.

Though brief and incomplete, the above discussion should have highlighted an established fact - the empirical proof that scale-invariant mechanisms are observed in the river basin. We refer, however, the reader to more

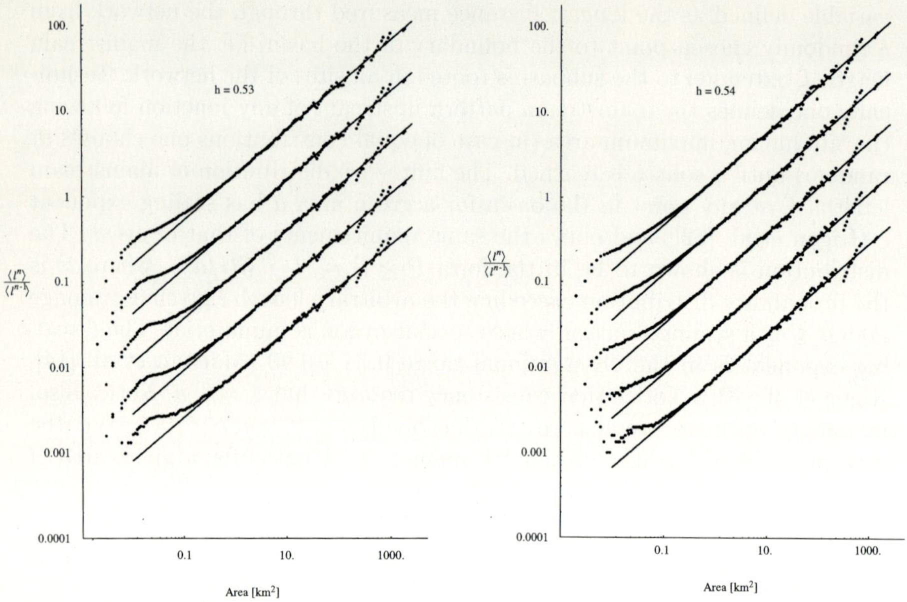

Fig. 9.3. Two double logarithmic plots of the scaling ratio $< l^n > / < l^{n-1} >$ vs area A for the Tug Fork and Boise river basins (after Rigon et al. [22]). Notice that the vertical scales are offset vertically to avoid overlapping of the curves referring to different moments. The plot shows a well defined simple scaling for at least three orders of magnitude. At very small scales a lower cutoff emerges owing to the transition from hillslope to fluvial settings. Notice that the parallel scalings of consecutive moments is strong observational evidence for the validity of the finite-size scaling ansatz (Maritan et a. [14]).

Exponent	Scheidegger trees	Leopold/Eden Networks	Real Basins
β	1/3	0.52 ± 0.02	0.43 ± 0.02
H	1/2	1	$0.75 - 0.80$
h	2/3	0.57	$0.57 - 0.60$
ξ	1	0.91 ± 0.05	$0.8 - 0.9$

Table 9.1. Summary Table of scaling coefficients for chance-dominated networks (Leopold/Eden and Scheidegger) compared with observational data (after Rodriguez-Iturbe and Rinaldo [28]).

complete accounts, also to render appropriately the rather involved matter of priorities (e.g. Rodriguez-Iturbe and Rinaldo [28]; Rinaldo et al. [25]).

9.4 Tidal Networks

Tidal networks (Figure 9.2) require new digital models to be automatically extracted from bathymetric fields (Fagherazzi et al. [4]), chiefly because topographic gradients are substituted by free-surface gradients in driving fluxes. As in the previous section we will not deal here with the objective tools to extract the networks' features from topography (see Fagherazzi et al. [4]).

Definite departures from the broad scaling features of river networks are observed. In particular, there seems to exist a much stronger sensitivity to environmental factors than in the rivers' case (Pestrong [20]; Knighton et al. [9]; Fagherazzi et al. [4]). Moreover the range of scales allowed by the processes is much narrower than that of fluvial networks. This per se restricts considerably the interest for scaling studies in the context of tidal networks. However, some scaling features of tidal networks (e.g. fat-fractal characters) are remarkable, whereas others, like those related to the aggregation structure, are of somewhat unclear definition (Fagherazzi et al. [4]). Also, questions arise that may potentially serve the broad scopes outlined in the introductory section: why do networks form in tidal environment? Why so few loops (i.e., paths following the network structure forcing one to pass through the same node more than once) are observed? Are tidal meanders similar to the fluvial ones?

In this section we address the issue of scale dependency or invariance in tidal network morphology, looking at the imprinting of competing processes on morphological scaling relationships and, through breaks of scale invariance, at mixed signatures of self-organizing processes. In particular, we introduce some important relationships related to watershed area, namely: the variations of drainage density within the tidal basin; Hack's relationship; the scaling of watershed area with channel width; the probability distributions for watershed areas, lengths and widths.

Drainage density is an important measure of the degree of dissection of a landscape by channelized patterns. In fluvial studies drainage density is traditionally defined as the ratio of total network length to watershed area, e.g. measured in m^{-1}; its inverse is proportional to the distance one has to walk, in the average, before encountering a channel. The concept of drainage density implies the existence of a fundamental length scale associated with the dissection of the landscape by the river network (Montgomery and Dietrich [17]) reflecting the efficiency of drainage and the fineness of the topographic texture. In tidal basins, however, the nonnegligible size of the channels with respect to the system size, and the varying degree of textures encountered, suggest the introduction of alternative measures of drainage density, reflecting both the typical flow length in unchanneled areas and the extent of channelized basin areas to total area. Here, following Rinaldo et al. [26], we define as a measure of the density of the channel network: a length, D_1 defined as the ratio of the watershed area divided by the total network length (i.e., the inverse of the drainage density as usually defined in fluvial studies); and a dimensionless ratio, D_2, of the number of channelized pixels within the basin divided by the total number of pixels making up the image, thereby defining objectively the occupation of area by the network extent. Figure 9.4 illustrates the results of the application of the above definitions, through the techniques to extract geometrical information described in Fagherazzi et al. [4], to four different tidal environments, respectively Petaluma marsh (CA), Barnstable (MA), a small salt marsh zone within the Venice lagoon (Pagliaga salt marsh) and the entire channelized part of the northern lagoon of Venice. In particular, the plots reported in Figure 9.4 have been obtained by defining for a large number of sections of the studied networks the relevant watershed (Rinaldo et al. [26]) and computing its drainage density parameters according to the above definitions. Technically, the data are discrete values obtained from the pixel units employed, i.e. one can only have integer multiples of the unit length (the pixel size) for a given area. The data organization depends on the computational procedure. In fact, when one computes D_1 for a branch of the network analyzed (say, from one bifurcation to the next downstream one), the features at the closure are assumed to be representative of the entire branch as they vary mildly and continuously. At junctions, area changes abruptly as well as the other features. An added complication, in the case of tidal network, is that, unlike rivers, effects of spatial morphological trends may play a nonnegligible role, i.e. quite different watersheds and embedded networks with equal total area might be located either close to the mouths or towards the inland zones.

Hack's [5] law is an empirical relationship that relates, within a river basin, total contributing area A to mainstream length, say L, computed by counting the number of pixels along the skeleton of the network by choosing the site with largest contributing area up to the watershed divide. In fluvial systems, Hack's law maintains a noteworthy consistency through many scales as a power-law of the type $L \propto A^{0.57 \pm 0.03}$. The significant departure from an exponent of 0.5, which would have characterized an Euclidean mathematical object (Mandelbrot [13]), has attracted considerable interest both from geomorphologists and in fractal studies, especially through finite-scale scaling arguments (e.g. Rodriguez-Iturbe and Rinaldo [28]). Figure 9.5 shows the area-mainstream length (log-log) plots for the same networks employed for the study of drainage densities (Figure 9.4). Mean values, i.e. ensemble averages of area given mainstream length, are also shown.

In fluvial basins it is commonly assumed that total contributing area A is proportional to landscape-forming discharges Q. This assumption is reasonable whenever the basin dimension does not exceed a scale characteristic of the heterogeneity of meaningful spatial patterns of intense rainfall, and this is commonly assumed to hold for several orders of magnitude. In tidal networks one faces the challenge of the interaction of ebb and flood flows in shaping the morphology of the network, and thus we wonder whether watershed areas have a morphodynamic meaning like their fluvial counterparts. In particular,

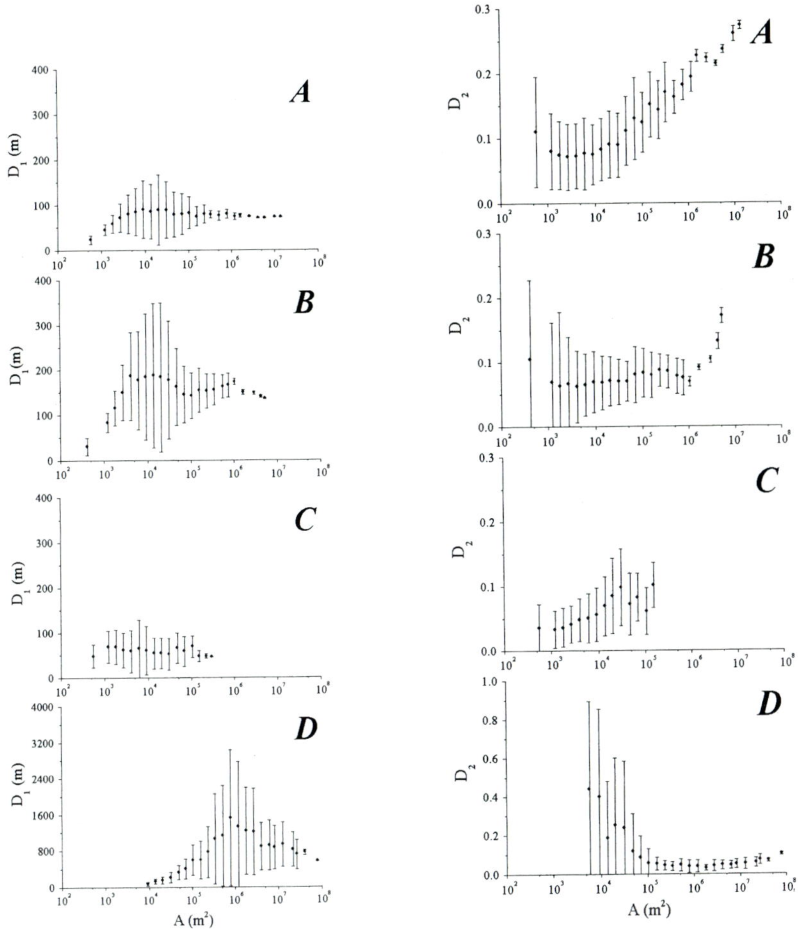

Fig. 9.4. Mean and variance of drainage density parameters D_1 and D_2 vs total watershed area A. The semi-logarithmic plot emphasizes the mean values (the expected value of the distribution inferred from the samples of drainage density given total watershed area) and the estimated variance. Network lengths needed to obtain the values of D_1 are computed along the skeleton of the patterns and measuring the total channelized lengths. (a) Petaluma Marsh (California); (b) Barnstable Marsh (Massachusetts); (c) Pagliaga salt marsh within the Venice Lagoon; (d) the entire channelized pattern of the Venice Lagoon (after Rinaldo et al. [26]).

the relationship between channel width and basin area is likely to be strongly affected by the very rapidly declining channel width typical of tidal estuaries (see references in Fagherazzi et al. [4]), and by the complex dynamics of lateral storage zones which ebb or flood only above an elevation threshold thus

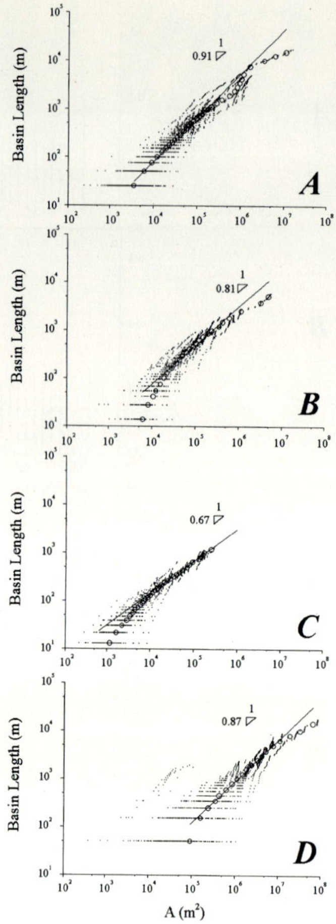

Fig. 9.5. Mainstream length of tidal subbasins vs total basin area for the same tidal environments as in Figure 9.4. Lengths are computed using the identification techniques described in Fagherazzi et al. [4]. The larger open circles represent the ensemble mean (mean drainage area given mainstream length). Slopes are fitted to the scaling of the mean values through selected portions of the graph whenever reasonable (after Rinaldo et al. [26]).

yielding strong nonlinear characters to the overall process. The relationship of watershed area A, computed via the procedure proposed in Rinaldo et al. [26], with the width w of the tidal channel at its closure, is shown in Figure 9.6.

The most important signatures of scale-invariant processes acting in the tidal basins are related to the existence of power-law characteristics in the tails of probability distributions of relevant geometrical quantities. We have

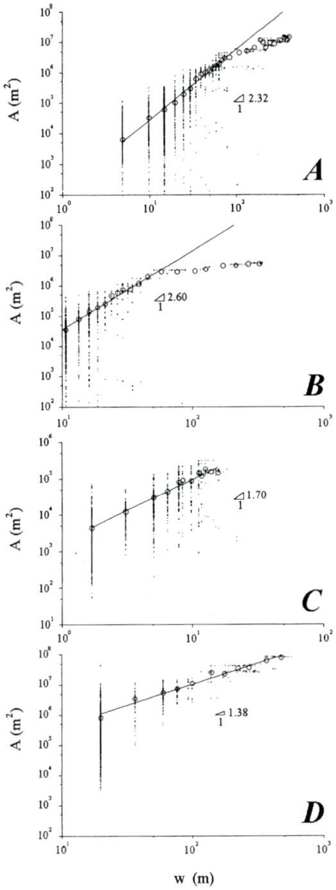

Fig. 9.6. Watershed area A vs cross-section width w computed for the four tidal environments of Figure 9.4 (Rinaldo et al. [26]).

studied three indicators, namely, the probability distributions for watershed areas, lengths and widths. Figure 9.7, a to d, shows a double logarithmic plot of the exceedence probability of watershed area, $P[\geq a]$ plotted against area a (here the upper case denotes the random variable and the lower case the current value, in square meters) for the four field cases, i.e. Petaluma marsh, Barnstable, the Pagliaga salt marsh and the northern part of the Venice lagoon. For the same environments, Figure 9.8 (a to d) shows the exceedence probability of total network length, measured upstream of each site. The latter is an interesting measure that, in river networks, defines Hortonian ratios of bifurcation and length (Rodriguez-Iturbe and Rinaldo [28]); Figure 9.9 a to d shows the exceedence probability of tidal channel cross-section

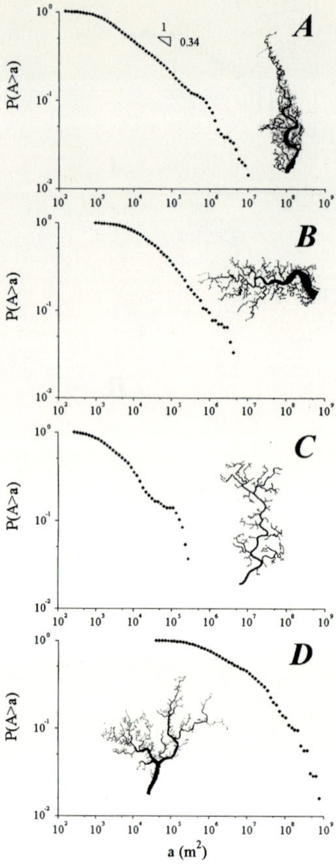

Fig. 9.7. Double logarithmic plot of the exceedence probability of watershed area $P[\geq a]$ (vs the current value of area a) computed for the usual four tidal environments, as in Figure 9.4 (after Rinaldo et al. [26]).

width, measured via the algorithm described in Fagherazzi et al. [4] and Rinaldo et al. [26]. Its statistics are important since width is functionally related to the landscape-forming flowrates.

9.5 Discussion

The data reported in the previous section suggest a large, site-dependent variability of the studied relationships reflecting the variety of physical and biological processes that may shape the tidal channels. We thus concur with the observation of Pethick [21] that the variability of drainage density (i.e., the inverse of D_1) within a tidal basin is one of the foremost differences

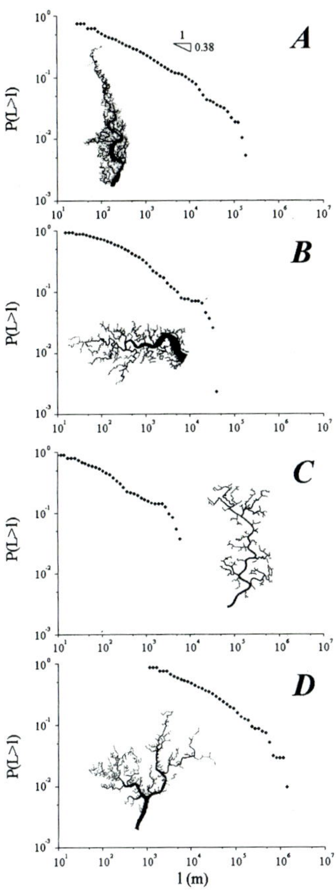

Fig. 9.8. Double logarithmic plot of the exceedence probability of lengths to the divide $P[\geq l]$ (vs the current value of length l) computed for the usual four tidal environments, as in Figure 9.4 (after Rinaldo et al. [26]).

with fluvial basins where, when homogeneous channel initiation mechanisms prevail, the drainage density ought to be constant at various scales (e.g. Montgomery and Dietrich [17]; Rodriguez-Iturbe and Rinaldo [28]). To the authors' knowledge, however, no systematic studies of the scaling of fluvial drainage densities, reliably identified in the field, have been produced so far. Furthermore, the dependence on area of the occupation of area by the tidal network, measured by D_2, is not simple but reflects heterogeneities of various nature in landscape-forming processes. At the scale of the basin, such heterogeities might also include sediment composition, although it is reasonable that the substrate of fluvial channels is much more variable than in the tidal cases. Much more likely, in our view, is a systematic emergence of crossovers

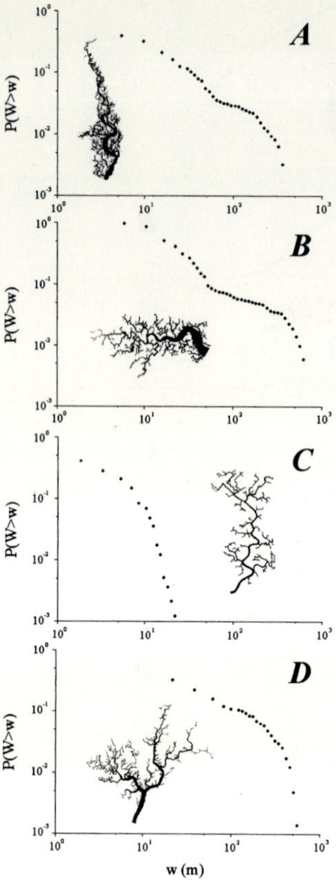

Fig. 9.9. Double logarithmic plot of the exceedence probability of channel width $P[\geq w]$ (vs the current value of width w) computed for the usual four tidal environments, as in Figure 9.4 (after Rinaldo et al. [26]).

(e.g., breaks in power law relationship) resulting from an overlapping between physical mechanisms acting at different space and/or time scales.

The heterogeneous character of the mechanisms controlling drainage density are also emphasized by the noteworthy variances shown in Figures 9.4 to 9.9. It is significant that the wide variation of morphological texture within the tidal basin can be easily detected, as in certain areas, in particular open to the larger tidal flows, the network tends to disappear towards a relatively amorphous, unchanneled flat. In such a case drainage density decreases quickly, (i.e. D_1 increases and D_2 decreases). In the case of the Venice lagoon, for instance, a few sites have unchanneled paths of the order of kilometers (Figure 9.4 d). We note, however, that only in two cases one observes for large areas a tendency to stabilize the values of D_1. The function D_2, that embeds the properties of finite transverse dimensions of the channels, shows a rather different behavior. We suggest that these features, reconsidered below in the framework of scaling areas and widths, might combine geomorphic signatures of the scaling properties of the channel widths and, possibly, of the transition from (seaward) flood-dominated areas (i.e., characterized by a shorter flood duration and higher flood current maximum)to (landward) ebb-dominated zones (i.e., characterized by a shorter fall period and greater ebb current maximum).

Also Hack's law (Figure 9.5) does not seem to apply for tidal networks, both because of the lack of clear power-law characters holding across a few orders of magnitude, or because scaling exponents, when at all suitable to estimation, differ considerably from the range seen in fluvial environments. It is apparent that the 'elongation' experienced by tidal subbasins, measured by the scaling exponent of length to areas, varies considerably, especially for small patterns, and that no common trend can be shown from our data which seem representative of a variety of possible tidal settings. The observations shown in Figure 9.5 (i.e., the reported scaling exponents shown mostly for visual comparison and to disprove the linearity assumption) differ considerably from the case of fluvial basins, which consistently elongate as they increase in size notwithstanding fluctuations, resulting in a robust scaling relationship with scaling exponents in the range $0.54 - 0.60$. These scaling exponents were argued to result from a finite-size scaling argument holding remarkably well in fluvial settings over several orders of magnitude (e.g. Rodriguez-Iturbe and Rinaldo [28]).

The trends in the Hack's plots for small areas may suggest that some tidal systems may become fluvially elongated (i.e., showing a power-law scaling of mainstream lengths and area across a significant range of lengths), thus probably ebb-dominated. In two cases (Figures 9.5 b and c) one might even suggest two different crossovers, roughly at the same area ($\sim 10^4$ m^2). In the log-linear cases, however, the scaling exponents are much larger than in the fluvial case. The flattening of the plots for very large areas, instead, most likely reflects morphological adjustments in the seaward, flood-dominated, regions.

What striking difference with rivers explains the failure of Hack's law and the rather complex patterns of drainage density for tidal basins? The interplay of variable erosional resistance; the unsteady and multidirectional landscape-forming flows (in very shallow depths it is reasonable that wind-induced wave resuspension is a leading gemorphic agent); variable sediment loadings (or lack of them); and significant time lags in morphological response, all taking place over relatively short spatial domains, could individually account for some of the differences. One basin parameter that might synthesize the emergence of a crossover of dynamic nature is the watershed area at which breaks in the plots occur. For Petaluma marsh, at approximately 10^4 m^2 both parameters related to drainage density show a break (Figure 9.4a), but rather surprisingly where one parameter is relatively constant, the other increases. Curiously, the only range in which a power-law makes sense in Hack's plot ($2 \times 10^3 - 10^5$ - Figure 9.5a) embodies exactly the range where the breaks occur for the drainage density parameters. Barnstable (Figures 9.4b and 9.5b) shows relatively constant drainage density in the range $\sim 2 \times 10^3 - 10^6$ m^2, and shows a break in Hack's plot around 5×10^4 m^2, like in the case of Petaluma, in a range where no crossover occurs in drainage density. A noticeable transition, in the case of the Lagoon of Venice (Figures 9.4c,d and 9.5c,d), occurs at area values $\sim 10^5$ m^2 for the area occupation by the network, but other than that it is difficult to see clear transitions, unless for Hack's plot of the smaller Pagliaga subset of the Lagoon (Figure 9.5c) for which two log-linear regimes are observed.

The watershed area versus channel width relationships may show rather sharp breaks. While there is a general tendency for a power-law relationship for all sites examined (Figure 9.6), we notice, however, that two of the sites (Figures 9.6a,b) show a strong break in the scaling, parting regimes where the exponents vary widely. Also, variances are large, suggesting a statistical rather than a deterministic framework for $A(w)$ relationships. The transition from one power-law regime to the next roughly corresponds to the break in Hack's plot (areas slightly larger that 10^6 m^2 for Petaluma and Barnstable).

Few of the plots of the relevant probability distributions (Figures 9.7-9.9) show scale-invariant power law distributions. Where reasonable, we have estimated the scaling coefficients from the log-log plots, and these are shown in the Figures. The larger environment (and, at the time of measure, less disturbed by human activity) is Petaluma marsh which shows signatures of scale invariance for three decades. Its area distribution shows an aggregation structure defined by the relationship $P[\geq a] \propto a^{-0.34}$ in a relatively large range of scales. The scaling exponent is significantly different from that usually shown by river basins (in the range $0.41 - 0.45$, as discussed in the previous section). Nearly all the distributions, however, show characteristic scales, i.e. a significant departure from the linear behavior. This is the case, in particular, of the tidal networks of the northern Lagoon of Venice, (Figures 9.7c,d to 9.9c,d), affected as they are today by the loss of a large part of the fine scale salt marsh network which presumably characterized its natural appearance. This is particularly evident in the surroundings of the larger channels, which have experienced a progressive increase in the forcing tidal flows. Interestingly, its smaller relatively intact salt marsh networks within the same lagoonal environment tends to a structure more similar to that of Barnstable or Petaluma.

Scale dependence has good dynamic reasons. For instance, from the planar structures obtained one clearly sees, at times, a local contraction of the channel width in the downstream direction which cannot be attributed to discharge decrease. Width often declines on big alluvial rivers as they experience a transition downstream from gravel bedded and weak banks to fine sand-bedded and strong banked. This and other dynamical processes, e.g. related to meandering, must be dealt with properly towards a general morphodynamic explanation. In any case, they preclude a simple relationship to hold in general between width and reference flow rates, possibly identified on a geomorphic basis.

9.6 Conclusions

We have compared selected observational results from the (relatively established) literature on river networks and that, much less refined, on the morphology of tidal networks extracted from digital maps. In particular, we have studied the possible emergence of scaling structures via important structural relationships, in principle applicable and defined both for the tidal basin and its networks and for the fluvial landscape: the variations of drainage density within the basin; Hack's relationship; the complex scaling of watershed area with channel width; the probability distributions for areas, lengths and widths. We believe that the sometimes consistent scales at which breaks in the structural relationships studied occur will prove useful information to drive field measurements and observations. Moreover, the comparison of similarities and departures among fluvial and tidal networks may provide the key to understand the dynamic mechanisms producing scale invariance in nature, and a tool for the validation of morphodynamic models. Whereas fluvial systems show various scale-invariant tendencies, there seems to be a nearly complete lack of such tendencies in tidal networks. We believe that an explanation of this clear observational fact will prove important to earth sciences, and possibly beyond.

References

1. Bak, P. (1996) *How Nature Works: the Science of Self-Organized Criticality* (New York: Copernicus-Springer).
2. Dietrich W.E., Wilson C.J., Montgomery D.R., McKean J, and Bauer R. (1992) Erosion thresholds and land surface morphology. *J. Geology* **20**, 675-679.
3. Eden, M. (1961) A two-dimensional growth process, Fourth Berkeley Symposium on Mathematical Statistics and Probability. In *Biology and Problems of Health*, ed. R Neyman, 4, 223-239. (Berkeley: Univ. of California Press).
4. Fagherazzi, S., Bortoluzzi, A., Adami, A., Marani, M., Lanzoni, S., and Rinaldo, A. (1999) Tidal Networks 1. Automatic extraction from digital terrain maps, *Water Resources Research*, 35, 3891-3904.
5. Hack, J.T. (1957) Studies of longitudinal profiles in Virginia and Maryland, U.S. Geological Survey Paper 294-B, Washington.
6. Howard, A.D. (1971) Simulation model of stream capture, *Geol. Soc. Am. Bulletin* **82**, 1355-1363.

7. Howard, A.D. (1994) Detachment-limited model of drainage basin evolution, *Water Resources Research* **30**, 2261-2285.
8. Kirchner, J.W. (1993) Statistical inevitability of Horton's laws and the apparent randomnness of stream channel networks, *Geology* **21**, 591-594.
9. Knighton, A.D., Woodroffe, C.D. and Mills, K. (1992) The evolution of tidal creek networks, Mary River, Northern Australia, *Earth Surface Processes and Landforms* **17**, 167-190.
10. Leopold, L.B., Langbein, W.B. (1962) The concept of entropy in landscape evolution, *U.S. Geological Survey Prof. Paper 500-A*, Washington.
11. Leopold L.B., Wolman MG, Miller J.P. (1964) *Fluvial Process in Geomorphology* (San Francisco: Freeman and Co.).
12. Mandelbrot, B.B. (1967) How long is the coast of Britain? Statistical self-similarity and fractional dimension, *Science* **156**, 636-638.
13. Mandelbrot, B.B. (1983) *The Fractal Geometry of Nature* (San Francisco: Freeman).
14. Maritan A, Rinaldo A, Giacometti A, Rigon R, and Rodriguez-Iturbe I. (1996) Scaling in river networks, *Physical Review E* **53**, 1501-1512.
15. McMahon, T.A. and Bonner, J.T. (1983) *On Size and Life* (New York: Scientific American Library).
16. Montgomery, D.R., and Dietrich W.E. (1988) Where do channels begin? , *Nature* **336**, 232-234.
17. Montgomery, D.R., and Dietrich W.E. (1992) Channel initiation and the problem of landscape scale, *Science* **255**, 826-830.
18. Murray, A.B., and Paola, C. (1994) A cellular model of braided rivers, *Nature* **371**, 54-57.
19. Niemeyer, L., Pietronero, L., and Wiesmann, H.J. (1984) Fractal Dimension of Dielectric Breakdown, *Physical Review Letters* **52**, 1033-1036.
20. Pestrong, R. (1965) The development of drainage patterns on tidal marshes, Stanford Univ. Publ. Geol. Sci., Tech. Rep. 10, 87 pp.
21. Pethick, J.S. (1992) Salt marsh geomorphology, In *Salt Marshes: Morphodynamics, Conservation and Engineering Significance*, ed. Allen, J.R.L. and Pye, K., Cambridge University Press, p. 184.
22. Rigon R., Rodriguez-Iturbe I., Giacometti A., Maritan A., Tarboton D., and Rinaldo, A. (1996) On Hack's law, *Water Resources Research* **32**, 3367-3374.
23. Rinaldo, A. Dietrich, W.E., Vogel, G., Rigon, R., and Rodriguez-Iturbe, I. (1995) Geomorphic signatures of varying climate, *Nature* **374**, 632-636.
24. Rinaldo, A., Maritan, A., Flammini, A., Colaiori, F., Rigon, R., and Rodriguez-Iturbe, I. (1996) Thermodynamics of Fractal Networks, *Physical Review Letters* **76**, 1501-1513.
25. Rinaldo, A., Rodriguez-Iturbe, I., and Rigon, R. (1998) Channel Networks, *Annual Review of Earth and Planetary Sciences* **26**, 289-327.
26. Rinaldo, A., S. Fagherazzi, Lanzoni, S., Marani, M., Dietrich, W.E. (1999) Tidal Networks 2. Watershed delineation and comparative network morphology, *Water Resour. Res.* **35**, 3905-3918.
27. Rodriguez-Iturbe I., Ijjasz-Vasquez E., Bras RL, and Tarboton D.G. (1992) Power-law distributions of mass and energy in river basins, *Water Resources Research* **28**, 988-993.
28. Rodriguez-Iturbe, I. and Rinaldo, A. (1997) *Fractal River Basins: Chance and Self-Organization* (New York: Cambridge University Press).

29. Sander, L.M. (1987) Fractal Growth, *Scientific American* **256**, 94-100.
30. Sapozhnikov, V.B. and Foufoula-Georgiou, E. (1998) Study of spatial scaling in braided river patterns using synthetic aperture radar imagery, *Water Resources Research* **34**, 1795-1804.
31. Scheidegger, A.E. (1967) *Theoretical Geomorphology*, 3rd edition (Berlin: Springer-Verlag).
32. Schumm, S.A. (1977) *The Fluvial System* (New York: J. Wiley)
33. Shreve, R.L. (1966) Statistical law of stream numbers, *J. Geology* **74**, 17-37.
34. Shreve, R.L. (1967) Infinite topologically random channel networks, *J. Geology* **77**, 397-414.

Printing (Computer to Film): Saladruck, Berlin
Binding: Stürtz AG, Würzburg